# BETA BEAMS

Neutrino Beams

# BETA BEAMS

## Neutrino Beams

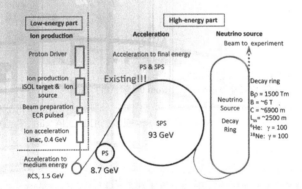

## Mats Lindroos
CERN, Switzerland

## Mauro Mezzetto
INFN, Sezione di Padova, Italy

**With a Foreword by Piero Zucchelli and**
**a Contribution on Low Energy Beta Beams by Cristina Volpe**

Imperial College Press

*Published by*

Imperial College Press
57 Shelton Street
Covent Garden
London WC2H 9HE

*Distributed by*

World Scientific Publishing Co. Pte. Ltd.
5 Toh Tuck Link, Singapore 596224
*USA office:* 27 Warren Street, Suite 401-402, Hackensack, NJ 07601
*UK office:* 57 Shelton Street, Covent Garden, London WC2H 9HE

**Library of Congress Cataloging-in-Publication Data**
Lindroos, Mats.
  Beta beams : neutrino beams / Mats Lindroos & Mauro Mezzetto.
  p. cm.
  Includes bibliographical references and index.
  ISBN 978-1-84816-377-5
  1. Neutrinos. 2. Neutron beams. I. Mezzetto, Mauro. II. Title.
  QC793.5.N425L56 2009
  539.7'375--dc22

                                    2009020410

**British Library Cataloguing-in-Publication Data**
A catalogue record for this book is available from the British Library.

Copyright © 2010 by Imperial College Press

*All rights reserved. This book, or parts thereof, may not be reproduced in any form or by any means, electronic or mechanical, including photocopying, recording or any information storage and retrieval system now known or to be invented, without written permission from the Publisher.*

For photocopying of material in this volume, please pay a copying fee through the Copyright Clearance Center, Inc., 222 Rosewood Drive, Danvers, MA 01923, USA. In this case permission to photocopy is not required from the publisher.

Printed in Singapore.

To our families.
Without the support from our wives, Rebecca and Maria-Cristina,
this book would not have been possible and more importantly,
without them life would not be worth living.

# Acknowledgments

Many thanks to all our colleagues in the world wide neutrino physics community. A special thanks to all our colleagues in the beta beam task group in the EC supported EURISOL Design Study (led by Michael Benedikt from CERN) and in the EC supported BENE network (led by Vittorio Palladino from the University of Naples and INFN) within the Infrastructure Activity CARE. Special thanks to Pierre Delahaye at GANIL for proof reading the Machine chapter. We would also like to thank John Ellis (CERN) for several precious suggestions.

The authors wish to thank their home institutes, CERN for Mats Lindroos and Istituto Nazionale di Fisica Nucleare for Mauro Mezzetto, for supporting this work. Mats Lindroos also wishes to thank the Physics Department at the University of Liverpool where he is Visiting Professor for their support.

We acknowledge the financial support of the European Community under the FP6 Research Infrastructure Action Structuring the European Research Area EURISOL DS Project Contract No. 515768 RIDS. The EC is not liable for any use that may be made of the information contained herein.

We acknowledge the support of the European Community-Research Infrastructure Activity under the FP6 Structuring the European Research Area programme (CARE, contract number RII3-CT-2003-506395).

# Contents

# Foreword

Innovation can achieve its target by either an incremental or radical change in knowledge. Neutrino physics is an outstanding field in this respect since, in less than 100 years, three elusive fundamental particles have been postulated, searched for, discovered, and characterized by giant steps requiring radical innovations. In almost all cases, the innovation to be undertaken was inconceivable by the same scientific community just a few years before. Well-known are the words of W. Pauli, the neutrino inventor: "I have done a terrible thing. I have postulated a particle that cannot be detected." And less than ten years later, F. Reines and C. Cowan were able to detect it, apparently driven by the fact that everybody said it could not be done. Not surprisingly, as F. Reines admitted "Clyne (Cowan) knew as little about the neutrino as I did but he was a good experimentalist with a sense of derring-do. So we shook hands and got off to working on neutrinos."

The beta beam is no more than a recent invention in the neutrino physics history: invention is the first occurrence of an idea for a new product or process, while innovation is the first attempt to carry it out into practice (J.E. Faberger, 2004). However, it will become a true innovation if it benefits from the potential synergy among different scientific disciplines. If you are not an insider in neutrino physics, this is your book: you will learn about neutrino physics in the broadest sense, from theory to acceleration and detection techniques, and you will have a very different view of the challenges and applications that are critical to the expansion of the beta beams potential.

All elements are apparently present to motivate your active contribution to beta beams: just a few years ago, many scientists insisted that they are conceptually impossible (still today, all artificial neutrino beams are produced by meson decay, as pioneered by Lederman, Schwartz and

Steinberger in the 60s). Therefore, don't forget your sense of derring-do, still necessary to pioneers of any time: if there are roses, they will blossom and the next book on beta beams will include your contribution.

Piero Zucchelli

# Chapter 1

# Introduction

## 1.1 Neutrino Oscillations

### 1.1.1 *Experiments*

The observation of neutrino oscillations has now established beyond doubt that neutrinos have mass and mix. This existence of neutrino masses is in fact the first solid experimental fact requiring physics beyond the Standard Model.

Since the early 1970s the chlorine solar neutrino experiment [1] provided evidence that electron neutrinos detected on earth were fewer than expected. In the 1980s this fact was confirmed by the water Čerenkov KamiokaNDE experiment [2] with a different detector threshold and the capability to demonstrate that the signals collected were indeed coming from the sun. A further experimental confirmation came in the early 1990s following two gallium experiments [3, 4] with very low detection thresholds. At the beginning of the 2000s, however, these experimental results were not considered evidence for neutrino oscillation, basically for two reasons: they were based on a comparison to theoretical predictions of the solar neutrino fluxes and multiple solutions could be found for the oscillation parameters, differing by several orders of magnitude to each other both in amplitude and in the neutrino mass difference squared.

This "solar neutrino puzzle" was closed in 2002 with the results from the SNO [5] and KamLAND [6] experiments. SNO, a heavy water Čerenkov solar neutrino detector, could simultaneously detect three solar neutrino processes: charged-current, elastic-scattering and neutral-currents, depending on different ratios of the $\nu_e$ and $\nu_\mu + \nu_\tau$ fluxes. In this way it was able to assess in a model-independent way that the total neutrino flux on earth ($\nu_e + \nu_\mu + \nu_\tau$) was as expected while the $\nu_e$ flux (the only neutrino flavor generated on the sun) was indeed depleted, in a way compatible with the previous solar neutrino experiments. The KamLAND experiment provided at the same time a measurement of the disappearance of electron antineutrinos from nuclear fission reactors in Japan and Korea. This established, in combination with the solar neutrino results, large mixing-angle (LMA) MSW oscillations as the solution for the solar neutrino oscillations together with a precise determination of the relevant neutrino mixing angle and of the corresponding mass difference squared (see Table 1.1).

The first evidence of neutrino oscillation came from atmospheric neutrinos. Since the late 80s there have been indications from atmospheric neutrino experiments that muon neutrinos disappear when going through the earth. This was finally unambiguously demonstrated by the Super-Kamiokande experiment in 1998 [7], a result well supported by the Soudan2 [8] and MACRO [9] experiments. This disappearance takes place at a much shorter wavelength than for solar neutrinos (L/E$\sim$ 500 km/GeV) [10]; it is not seen for electron neutrinos, a fact that has been best established by the CHOOZ reactor experiment [11]. This disappearance has been confirmed by two long-baseline experiments: the K2K experiment in Japan [12] and MINOS in USA [13] (see Section 1.3.2).

A further, much more controversial, indication of $\bar{\nu}_\mu \rightarrow \bar{\nu}_e$ oscillations with a $\Delta m^2$ of 0.3 - 20 eV$^2$ came from the beam dump LSND experiment detecting a $\sim 4\sigma$ excess of $\bar{\nu}_e$ interactions in a neutrino beam produced by $\pi^+$ decays at rest where the $\bar{\nu}_e$ component was highly suppressed ($\sim 7.8 \cdot 10^{-4}$) [14]. The KARMEN experiment [15], using a very similar technique but with a lower sensitivity (a factor 10 less for the lower $\Delta m^2$), and the NOMAD experiment at WANF of CERN SPS [16], for $\Delta m^2 > 10$ eV$^2$, did not confirm the result, excluding a large part of the allowed region of the oscillation parameters. The results of the MiniBooNE experiment [17] again did not confirm the LSND result, even though some non-standard explanations, for instance sterile neutrinos [18] or $\nu_e$ disappearance [19], have not been fully excluded.

### 1.1.2  *Phenomenology*

The above experimental observations are consistently described by three families $\nu_1, \nu_2, \nu_3$ with mass values $m_1$, $m_2$ and $m_3$ that are connected to the flavor eigenstates $\nu_e$, $\nu_\mu$ and $\nu_\tau$ by a mixing matrix $U$, usually parameterized as

$$U(\theta_{12}, \theta_{23}, \theta_{13}, \delta_{\rm CP}) =$$

$$\begin{pmatrix} c_{13}c_{12} & c_{13}s_{12} & s_{13}e^{-i\delta_{\rm CP}} \\ -c_{23}s_{12} - s_{13}s_{23}c_{12}e^{i\delta_{\rm CP}} & c_{23}c_{12} - s_{13}s_{23}s_{12}e^{i\delta_{\rm CP}} & c_{13}s_{23} \\ s_{23}s_{12} - s_{13}c_{23}c_{12}e^{i\delta_{\rm CP}} & -s_{23}c_{12} - s_{13}c_{23}s_{12}e^{i\delta_{\rm CP}} & c_{13}c_{13} \end{pmatrix} \quad (1.1)$$

where the short-form notation $s_{ij} \equiv \sin\theta_{ij}, c_{ij} \equiv \cos\theta_{ij}$ is used. As a result, the neutrino oscillation probability depends on three mixing angles, $\theta_{12}, \theta_{23}, \theta_{13}$, two mass differences, $\Delta m_{12}^2 = m_2^2 - m_1^2$, $\Delta m_{23}^2 = m_3^2 - m_2^2$, and a CP phase $\delta_{\rm CP}$. Additional phases are present in case neutrinos are Majorana particles, but they do not influence neutrino flavor oscillations at all. Furthermore, the neutrino mass hierarchy, the order by which mass eigenstates are coupled to flavor eigenstates, can be fixed by measuring the sign of $\Delta m_{23}^2$. In vacuum the oscillation probability between two neutrino flavors $\alpha$, $\beta$ is:

$$P(\nu_\alpha \to \nu_\beta) = -4 \sum_{k>j} Re[W_{\alpha\beta}^{jk}] \sin^2 \frac{\Delta m_{jk}^2 L}{4E_\nu} \pm 2 \sum_{k>j} Im[W_{\alpha\beta}^{jk}] \sin^2 \frac{\Delta m_{jk}^2 L}{2E_\nu},$$
$$(1.2)$$

where $\alpha = e, \mu, \tau$, $j = 1, 2, 3$, $W_{\alpha\beta}^{jk} = U_{\alpha j} U_{\beta j}^* U_{\alpha k}^* U_{\beta k}$. In the case of only two neutrino flavor oscillations it can be written as:

$$P(\nu_\alpha \to \nu_\beta) = \sin^2 2\theta \cdot \sin^2 \frac{1.27 \, \Delta m^2(eV^2) \cdot L(km)}{E_\nu(GeV)}. \quad (1.3)$$

When neutrinos pass through matter, the oscillation probability is perturbed [20] (see Eq. 1.5 in the following section). Two independent mass splittings characterize the system, since oscillations only depend on the difference of squared masses. Although no formally agreed definition exists, the usage is that the mass eigenstates are classified by decreasing electron-neutrino content $| < \nu_e|\nu_1 > |^2$, $| < \nu_\mu|\nu_2 > |^2$, $| < \nu_\tau|\nu_3 > |^2$. With this definition, the mass of $\nu_1$ is not necessarily smaller than that of $\nu_2$.

Since two-family neutrino oscillations in vacuum (Eq. 1.3) depend on the mass difference as $\sin^2(1.27\Delta m^2 L/E)$, one cannot determine the sign of $\Delta m^2$ unless the oscillation interferes with another process; in the case of electron neutrinos, this is offered by coherent scattering on electrons in matter, a.k.a. matter effects.

The best-fit values and allowed range of values of the oscillation parameters at different CL obtained in [22] are shown in Table 1.1.

Table 1.1  Best-fit values, $2\sigma$, and $3\sigma$ intervals (1 dof) for the three flavor neutrino oscillation parameters from global data including solar, atmospheric, reactor (KamLAND and CHOOZ) and accelerator (K2K and MINOS) experiments [22].

| parameter | best-fit | $2\sigma$ | $3\sigma$ |
|---|---|---|---|
| $\Delta m_{21}^2\,[10^{-5}\text{eV}^2]$ | $7.65^{+0.23}_{-0.20}$ | 7.25–8.11 | 7.05–8.34 |
| $|\Delta m_{31}^2|\,[10^{-3}\text{eV}^2]$ | $2.40^{+0.12}_{-0.11}$ | 2.18–2.64 | 2.07–2.75 |
| $\sin^2\theta_{12}$ | $0.304^{+0.022}_{-0.016}$ | 0.27–0.35 | 0.25–0.37 |
| $\sin^2\theta_{23}$ | $0.50^{+0.07}_{-0.06}$ | 0.39–0.63 | 0.36–0.67 |
| $\sin^2\theta_{13}$ | $0.01^{+0.016}_{-0.011}$ | $\leq 0.040$ | $\leq 0.056$ |

## 1.2  Three-family Oscillations and CP or T Violation

Three parameters have not yet been measured in neutrino oscillations: $\theta_{13}$, $\text{sign}(\Delta m_{23}^2)$ and $\delta_{\text{CP}}$, all three fundamental parameters of the standard model.

The mixing angle $\theta_{13}$ is the key parameter of three-neutrino oscillations and regulates at the first order all the oscillation processes that could contribute to the measurement of $\text{sign}(\Delta m_{23}^2)$ and $\delta_{\text{CP}}$.

The $\text{sign}(\Delta m_{23}^2)$ parameter is an internal degree of freedom in the neutrino sector; its value could be $+1$ (normal hierarchy), in which case $\nu_e$ would be the lightest neutrino, or $-1$ (inverted hierarchy), for which $\nu_e$ would be the heaviest (see Fig. 1.1). Its value is of great importance for double-beta decay experiments [23] and for great unification model building.

The CP phase $\delta_{\text{CP}}$ is the holy grail of ultimate neutrino oscillation searches. The demonstration of CP violation in the lepton sector (LCPV) and the knowledge of the value of this phase would be crucial to understanding the origin of the baryon asymmetry in the universe, providing a strong indication, though not proof, that leptogenesis is the explanation for the observed baryon asymmetry of the Universe [24].

All these parameters can be measured via subleading $\nu_\mu \to \nu_e$ oscillations that represent the key process of any future new discovery in neutrino oscillation physics.

The importance of precision measurements of the already measured neutrino parameters should not be underestimated. The atmospheric

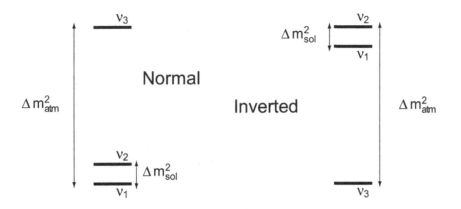

Fig. 1.1   The two three-neutrino mass schemes, normal and inverted hierarchy.

parameters can be precisely measured by long-baseline neutrino oscillation experiments via the $\nu_\mu \to \nu_\mu$ oscillation process.

### 1.2.1   *How to measure leptonic CP violation*

After the discovery of neutrino oscillations in 1998, it was soon realized that with three families and for a favorable set of parameters, it would be possible to observe violation of CP or T symmetries in neutrino oscillations [25]. This observation reinforced the considerable interest for precision measurements of neutrino oscillation parameters.

The year 2002 was very encouraging for LCPV projects, since the LMA solution emerged as the true solution for solar neutrino oscillations. Only for this solution can leptonic CP violation be large enough to be observed in high-energy neutrino oscillation appearance experiments.

This has led to extensive studies, such as those published in a CERN yellow report [26], the European Network BENE [27] or the International Scoping Study [28].

The phenomenon of CP (or T) violation in neutrino oscillations manifests itself by a difference in the oscillation probabilities of say, $P(\nu_\mu \to \nu_e)$ vs $P(\overline{\nu}_\mu \to \overline{\nu}_e)$ (CP violation), or $P(\nu_\mu \to \nu_e)$ vs $P(\nu_e \to \nu_\mu)$ (T violation).

It can be observed right away that observation of this important phenomenon requires appearance experiments; indeed a reactor or solar neutrino experiment, sensitive to the disappearance $P(\nu_e \to \nu_e)$ which is clearly time-reversal invariant, would be completely insensitive to it. This can be seen as an advantage in view of a precise and unambiguous measurement of the mixing angles; for the long-term goal of observing and studying CP violation, we are confined to appearance experiments.

The $\nu_\mu \to \nu_e$ transition probability in case of small matter effects can be parameterized as [29]:

$$
\begin{aligned}
P(\nu_\mu \to \nu_e) &= 4c_{13}^2 s_{13}^2 s_{23}^2 \sin^2 \frac{\Delta m_{13}^2 L}{4E_\nu} \times \left[ 1 \pm \frac{2a}{\Delta m_{13}^2}(1 - 2s_{13}^2) \right] \\
&+ 8c_{13}^2 s_{12} s_{13} s_{23}(c_{12}c_{23}\cos\delta_{\mathrm{CP}} - s_{12}s_{13}s_{23}) \cos \frac{\Delta m_{23}^2 L}{4E_\nu} \sin \frac{\Delta m_{13}^2 L}{4E_\nu} \sin \frac{\Delta m_{12}^2 L}{4E_\nu} \\
&\mp 8c_{13}^2 c_{12}c_{23}s_{12}s_{13}s_{23}\sin\delta_{\mathrm{CP}} \sin \frac{\Delta m_{23}^2 L}{4E_\nu} \sin \frac{\Delta m_{13}^2 L}{4E_\nu} \sin \frac{\Delta m_{12}^2 L}{4E_\nu} \\
&+ 4s_{12}^2 c_{13}^2 \{c_{13}^2 c_{23}^2 + s_{12}^2 s_{23}^2 s_{13}^2 - 2c_{12}c_{23}s_{12}s_{23}s_{13}\cos\delta_{\mathrm{CP}}\} \sin \frac{\Delta m_{12}^2 L}{4E_\nu} \\
&\mp 8c_{12}^2 s_{13}^2 s_{23}^2 \cos \frac{\Delta m_{23}^2 L}{4E_\nu} \sin \frac{\Delta m_{13}^2 L}{4E_\nu} \frac{aL}{4E_\nu}(1 - 2s_{13}^2).
\end{aligned}
\tag{1.4}
$$

The first line of this parameterization contains the term driven by $\theta_{13}$, the second and third contain CP even and odd terms respectively, and the fourth is driven by the solar parameters. The last line parameterizes matter effects developed at the first order where $a[\mathrm{eV}^2] = \pm 2\sqrt{2}G_F n_e E_\nu = 7.6 \cdot 10^{-5}\rho[g/cm^3]E_\nu[\mathrm{GeV}]$. The $\pm$ and $\mp$ terms refer to neutrinos and antineutrinos. A sketch of $P(\nu_\mu \to \nu_e)$ as a function of $L$ for 1 GeV neutrinos is shown in Fig. 1.2.

When matter effects are not negligible, following Eq. (1) of [32], the transition probability $\nu_e \to \nu_\mu$ ($\bar{\nu}_e \to \bar{\nu}_\mu$) at second order in perturbation theory in $\theta_{13}$, $\Delta m_{12}^2/\Delta m_{23}^2$, $|\Delta m_{12}^2/a|$ and $\Delta m_{12}^2 L/E_\nu$ (see also [20]) is:

$$
P^\pm(\nu_e \to \nu_\mu) = X_\pm \sin^2(2\theta_{13}) + Y_\pm \cos(\theta_{13})\sin(2\theta_{13})\cos\left( \pm\delta - \frac{\Delta m_{23}^2 L}{4E_\nu} \right) + Z,
\tag{1.5}
$$

where $\pm$ refers to neutrinos and antineutrinos, respectively. The coefficients of the two equations are:

$$
\begin{cases}
X_\pm = \sin^2(\theta_{23})\left( \frac{\Delta m_{23}^2}{|a - \Delta m_{23}^2|} \right)^2 \sin^2\left( \frac{|a - \Delta m_{23}^2|L}{4E_\nu} \right), \\[2ex]
Y_\pm = \sin(2\theta_{12})\sin(2\theta_{23})\left( \frac{\Delta m_{12}^2}{a} \right)\left( \frac{\Delta m_{23}^2}{|a - \Delta m_{23}^2|} \right)\sin\left( \frac{aL}{4E_\nu} \right)\sin\left( \frac{|a - \Delta m_{23}^2|L}{4E_\nu} \right), \\[2ex]
Z = \cos^2(\theta_{23})\sin^2(2\theta_{12})\left( \frac{\Delta m_{12}^2}{a} \right)^2 \sin^2\left( \frac{aL}{4E_\nu} \right)
\end{cases}
\tag{1.6}
$$

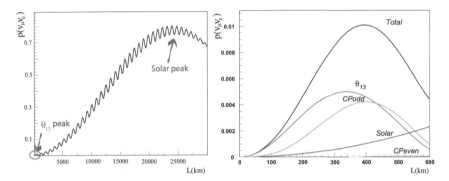

Fig. 1.2 Sketch of $P(\nu_\mu \to \nu_e)$ as function of the baseline computed for monochromatic neutrinos of 1 GeV in the solar baseline regime for $\delta_{CP}=0$ (left) and in the atmospheric baseline regime for $\delta_{CP} = -\pi/2$ (right), where the different terms of Eq. (1.4) are displayed. The following oscillation parameters were used in both cases: $\sin^2 2\theta_{13} = 0.01$, $\sin^2 2\theta_{12} = 0.8$, $\Delta m_{23}^2 = 2.5 \cdot 10^{-3}$ eV$^2$, $\Delta m_{12}^2 = 7 \cdot 10^{-5}$ eV$^2$. From Ref. [30].

Table 1.2 The 90%($3\sigma$) bounds (1 dof) on $\sin^2 \theta_{13}$ from an analysis of different sets of data read as [22]

$$\sin^2 \theta_{13} \leq \begin{cases} 0.060 \ (0.089) & \text{(solar+KamLAND)} \\ 0.027 \ (0.058) & \text{(CHOOZ+atm+K2K+MINOS)} \\ 0.035 \ (0.056) & \text{(global data)} \end{cases}$$

(remember that $a$ changes sign by changing neutrinos with antineutrinos and that $P(\nu_e \to \mu_\mu, \delta_{CP}) = P(\nu_\mu \to \nu_e, -\delta_{CP})$).

$\theta_{13}$ searches look for experimental evidence of $\nu_e$ appearance in excess of what is expected from the solar terms. These measurements will be experimentally hard because the present limit on $\theta_{13}$, summarized in Table 1.2, translates into a $\nu_\mu \to \nu_e$ appearance probability much smaller than 10% at the appearance maximum in a high energy muon neutrino beam.

One of the interesting aspects of Eq. (1.5) is the occurrence of matter effects which, unlike the straightforward $\theta_{13}$ term, depends on the sign of the mass difference sign($\Delta m_{23}^2$). These terms should allow extraction of the mass hierarchy, but could also be seen as a background to the CP violating effect, from which they can be distinguished by the very different neutrino energy dependence, matter effects being larger for higher energies, with a "matter resonance" at about 12 GeV.

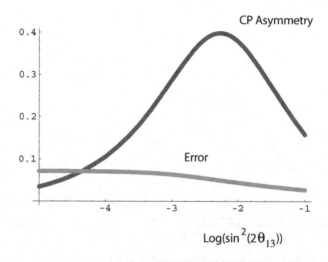

Fig. 1.3   Magnitude of the CP asymmetry at the first oscillation maximum, for $\delta = 1$ as a function of the mixing angle $\sin^2 2\theta_{13}$. The curve marked "error" indicates the dependence of the statistical+systematic error on such a measurement. The curves have been computed for the baseline beta beam option at the fixed energy $E_\nu = 0.4$ GeV, $L = 130$ km, statistical + 2% systematic errors. From [31].

The CP violation can be seen as interference between the solar and atmospheric oscillation for the same transition. Of experimental interest is the CP-violating asymmetry $A_{CP}$:

$$A_{CP} = \frac{P(\nu_\mu \rightarrow \nu_e) - P(\overline{\nu}_\mu \rightarrow \overline{\nu}_e)}{P(\nu_\mu \rightarrow \nu_e) + P(\overline{\nu}_\mu \rightarrow \overline{\nu}_e)} \qquad (1.7)$$

displayed in Fig. 1.3, or the equivalent time reversal asymmetry $A_T$.

The asymmetry can be large and its value increases for decreasing values of $\theta_{13}$ until the two oscillations (solar and atmospheric) are of the same magnitude [31]. The following remarks can be made:

(1) Contrary to naive expectations, the most favorable $\theta_{13}$ values for LCPV searches are not the highest allowed.
(2) This asymmetry is valid for the first maximum. At the second oscillation maximum the curve is shifted to higher values of $\theta_{13}$ so that it can be then an interesting possibility for measuring the CP asymmetry, although the reduction in flux is considerable (roughly factor 9).
(3) The asymmetry has opposite signs for $\nu_e \rightarrow \nu_\mu$ and $\nu_e \rightarrow \nu_\tau$, which changes when going from one oscillation maximum to the next.
(4) The asymmetry is small for large values of $\theta_{13}$, placing a challenging emphasis on systematics.

### 1.2.2 *The problem of degenerate solutions*

The richness of the $\nu_\mu \to \nu_e$ transition is also its weakness: it will be very difficult for pioneering experiments to extract all the genuine parameters unambiguously. Due to the three-flavor structure of the oscillation probabilities, for a given experiment several different disconnected regions of the multi-dimensional space of parameters could fit the experimental data, originating degenerate solutions.

Traditionally these degeneracies are referred to in the following ways:

- The *intrinsic* or $(\delta_{\rm CP}, \theta_{13})$-degeneracy [32]: for a measurement based on the $\nu_\mu \to \nu_e$ oscillation probability for neutrinos and antineutrinos, two disconnected solutions appear in the $(\delta_{\rm CP}, \theta_{13})$ plane.
- The *hierarchy* or $\text{sign}(\Delta m_{31}^2)$-degeneracy [33]: the two solutions corresponding to the two signs of $\Delta m_{31}^2$ appear in general at different values of $\delta_{\rm CP}$ and $\theta_{13}$.
- The *octant* or $\theta_{23}$-degeneracy [34]: since LBL experiments are sensitive mainly to $\sin^2 2\theta_{23}$ it is difficult to distinguish between the two octants $\theta_{23} < \pi/4$ and $\theta_{23} > \pi/4$. Again, the solutions corresponding to $\theta_{23}$ and $\pi/2 - \theta_{23}$ appear in general at different values of $\delta_{\rm CP}$ and $\theta_{13}$.

This leads to an eight-fold ambiguity in $\theta_{13}$ and $\delta_{\rm CP}$ [35], and hence degeneracies provide a serious limitation for the determination of $\theta_{13}$, $\delta_{\rm CP}$, and the sign of $\Delta m_{31}^2$. Discussions of degeneracies can be found for example in Refs. [36–39]; degeneracies in the context of the CERN–Fréjus beta beam and SPL have been considered for the first time in Ref. [40]. Degeneracies in the beta beam context are discussed in Section 3.5.2.

## 1.3 Experimental Setups

### 1.3.1 *Conventional neutrino beams*

Conventional neutrino beams are produced through the decay of $\pi$ and K mesons generated by a high energy proton beam hitting small Z, needle-shaped, segmented targets. Positive (negative) mesons are sign-selected and focused (defocused) by large acceptance magnetic lenses into a long evacuated decay tunnel where $\nu_\mu$'s ($\overline{\nu}_\mu$'s) are generated.

The length of the decay tunnel has to be optimized in order to maximize pion decays while keeping low the rate of muon decays, which contribute to the intrinsic $\nu_e$ beam contamination. Downstream the decay tunnel a beam

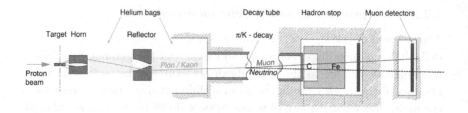

Fig. 1.4   Sketch of the CNGS neutrino beam line.

dump stops all the charged mesons while high energy muons are stopped downstream the beam dump by the earth: only neutrinos are allowed to reach the detector.

A not negligible fraction of the neutrinos hitting the detector is produced by secondary interactions in the target or in the material downstream. In case of positive charge selection, the $\nu_\mu$ beam has typically a few percent of $\bar{\nu}_\mu$ contamination (from the decay of the residual $\pi^-, K^-$ and $K^0$) and $\sim 1\%$ of $\nu_e$ and $\bar{\nu}_e$ coming from three-body $K^\pm$, $K_0$ decays and $\mu$ decays.

The precision of the evaluation of the intrinsic $\nu_e$ to $\nu_\mu$ contamination is limited by the knowledge of the $\pi$ and $K$ production in the primary proton beam target. Hadroproduction measurements at 400 and 450 GeV/c performed with the NA20 [41] and SPY [42] experiments at the CERN SPS provided results with $5 \div 7\%$ intrinsic systematic uncertainties. The Harp experiment [43] measured both the K2K [44] and the MiniBooNE [45] targets, covering most of the useful pion phase-space, successfully improving the description of the two beam lines.

Close detectors are used to directly measure beam neutrinos and backgrounds this issue will be further discussed in Section 1.4.

### 1.3.2   *First generation long-baseline experiments*

The first generation of long-baseline (LBL) experiments focused on confirming the atmospheric evidence of oscillations and measuring $\sin^2 2\theta_{23}$ and $|\Delta m_{23}^2|$ within $10 \div 15\%$ of accuracy if $|\Delta m_{23}^2| > 10^{-3}$ eV$^2$. It was pioneered by the K2K experiment at KEK [12], which ended data taking in 2004 and confirmed the SuperKamiokande atmospheric oscillation evidence at $4.3\sigma$.

K2K had a baseline of 250 km, and the muon-neutrino average energy was 1.2 GeV. The beam was created from a 12 GeV proton beam, 0.015 MW beam power. A dedicated close detector complex, with a 1 kt water Čerenkov tank, fine-grained detectors, and a muon ranger, was located 100 m from the end of the pion-decay volume. Super-Kamiokande was used as the far detector, and the first beam-induced neutrino event was observed in the summer of 1999. The final oscillation analysis [46] was performed using a data set corresponding to $0.922 \times 10^{20}$ protons on target. Were observed 112 beam-originated neutrino events, where the expected number in the absence of oscillations was $158.1^{+9.2}_{-8.6}$. Of these events, 58 were single-ring muon-like events fully contained within the Super-Kamiokande detector. The energies and directions of the muons in fully contained events can be reconstructed, and because of the simple kinematics of the charged-current quasi-elastic (CCQE) events that make up much of the cross section around 1 GeV, it is possible to estimate the energy of the incoming neutrinos. Spectral distortions were in good agreement with the expectations derived from the neutrino disappearance. These results support maximal mixing, with best-fit two-neutrino oscillation parameters of $\sin^2 2\theta = 1$ and $\Delta m^2 = 2.8 \times 10^{-3} \text{eV}^2$. The 90% CL range for $\Delta m^2$ at $\sin^2 2\theta = 1$ is between 1.9 and 3.5 $\times 10^{-3} \text{eV}^2$.

The MINOS (Main Injector Neutrino Oscillation Search) experiment was also proposed in 1995, with a neutrino beam, NuMI[47], pointed from Fermilab to the Soudan mine in Minnesota, with a baseline of 735 km. The beam has a system of movable focusing horns to allow the beam energy spectrum to be altered. Both near and far detectors consist of a steel and plastic scintillator sandwich structure.

The experiment started running in the spring of 2005, and within a year had gathered data corresponding to $2.50 \times 10^{20}$ protons on target. The MINOS results support maximal mixing, with best-fit parameters of $|\Delta m^2| = (2.43 \pm 0.13) \times 10^{-3} \text{eV}^2$ and $\sin^2(2\theta) > 0.95$ at 68% confidence level [48].

The oscillation parameters from the K2K and MINOS experiments, together with results from SuperKamiokande are shown in Fig. 1.5. MINOS will run for five years, with the goal of accumulating $16 \times 10^{20}$ protons on target. This data set should improve our knowledge of the oscillation parameters substantially. Both the experiments described here are linked, if only indirectly, to future projects to make precision measurements of the oscillation parameters and to probe the third mixing angle.

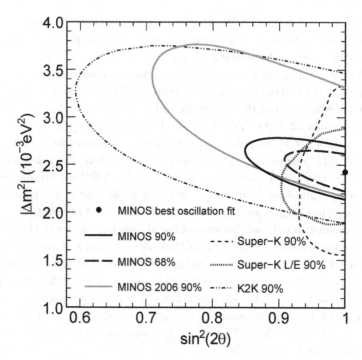

Fig. 1.5   Confidence intervals of atmospheric parameters as measured by MINOS, K2K and SuperKamiokande. From [48].

ICARUS [49] and OPERA [50] at the CNGS beam [51] from CERN to Gran Sasso laboratories will search for evidence of $\nu_\tau$ interactions in a $\nu_\mu$ beam, the final proof of $\nu_\mu \to \nu_\tau$ oscillations. CNGS had a first run in the 2006, while the OPERA detector started data taking with the full detector in 2008.

The CNGS $\nu_\mu$ beam has been optimized for the $\nu_\mu \to \nu_\tau$ appearance search in order to overcome the kinematic threshold for $\tau$ production and to detect the $\tau$ decay products. The average neutrino energy, 17 GeV, is about ten times higher than the optimal value for $\theta_{13}$ searches at the CERN-LNGS baseline of 732 km.

Current long-baseline experiments with conventional neutrino beams can look for $\nu_\mu \to \nu_e$ oscillations (see Section 2) even if they are not optimized for such studies. MINOS at NuMI is expected to reach a sensitivity of $\sin^2 2\theta_{13} = 0.08$ [13] integrating $14 \cdot 10^{20}$ protons on target (pot) in five years according to the FNAL proton plan evolution [52]. MINOS's main limitation is the poor electron identification efficiency of the detector.

Thanks to the dense ECC structure and the high granularity provided by the nuclear emulsions, the OPERA detector is also suited for electron detection [50]. OPERA can reach a 90% CL sensitivity of $\sin^2 2\theta_{13} = 0.06$ ($\Delta m_{23}^2 = 2.5 \cdot 10^{-3}$ eV$^2$, convoluted to CP and matter effects) [53], a factor $\sim 2$ better than CHOOZ for five years exposure to the CNGS beam at nominal intensity of $4.5 \cdot 10^{19}$ pot/yr.

### 1.3.3 *Second generation long-baseline experiments*

The focus of second generation LBL experiments will be the measurement of $\theta_{13}$ through the detection of sub-leading $\nu_\mu \rightarrow \nu_e$ oscillations.

According to the present experimental situation, conventional neutrino beams can be improved and optimized for the $\nu_\mu \rightarrow \nu_e$ searches. The design of a such new facility will demand higher neutrino fluxes, a neutrino beam optimized to the atmospheric $\Delta m_{23}^2$ and a detector optimized to efficiently detect electrons and reject $\pi^\circ$'s.

An interesting option for neutrino beams is the possibility to tilt the beam axis a few degrees with respect to the position of the far detector (off-axis beams) [54, 55]. According to the two-body $\pi$-decay kinematics, all the pions above a given momentum produce neutrinos of similar energy at a given angle $\theta \neq 0$ with respect to the direction of the parent pion (contrary to the $\theta = 0$ case where the neutrino energy is proportional to the pion momentum).

These neutrino beams have several advantages with respect to the corresponding on-axis ones: they are narrower, lower-energy and with a smaller $\nu_e$ contamination (since $\nu_e$ mainly come from three-body decays) although the neutrino flux can be significantly smaller.

In the next section the major players of this activity will be described, a sketch of $\theta_{13}$ sensitivities as a function of the time, following the schedule reported in the experimental proposals, is reported in Fig. 1.6.

#### 1.3.3.1 *T2K*

The T2K (Tokai to Kamioka) experiment [54] will aim neutrinos from the Tokai site of J-PARC (50 GeV, 0.75 MW) to the Super-Kamiokande detector 295 km away. The neutrino beam is situated at an off-axis angle of 2.5 degrees, ensuring a pion decay peak energy of about 0.6 GeV. The beam line is equipped with a set of dedicated on-axis and off-axis near detectors

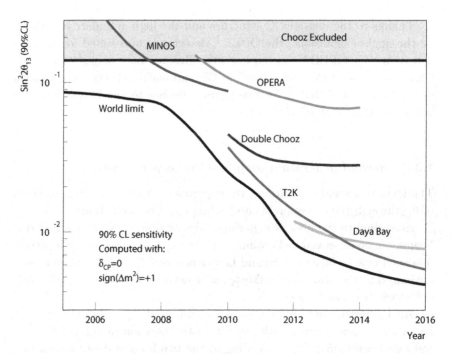

Fig. 1.6  Evolution of experimental $\sin^2 2\theta_{13}$ sensitivities as function of time. T2K sensitivity curve has been computed with the beam intensity curve of [56], Nova sensitivity is computed for 25 kt fiducial volume, $6.5 \cdot 10^{20}$ pot/yr, starting in 2012. Experiments are assumed to provide results after the first year of data taking.

at a distance of 280 m. In particular the off-axis near detector (ND280) will consist of finely segmented detectors acting as neutrino targets and tracking detectors surrounded by a magnet (recuperated from the UA1-NOMAD magnet at CERN). The purpose of the detector is to measure the neutrino spectrum, flux, $\nu_e$ contamination and interaction cross sections before the oscillation.

It is expected that the sensitivity of the experiment in a five-year $\nu_\mu$ run, designed to begin on April 1, 2009, will be of the order of $\sin^2 2\theta_{13} \leq 0.006$ (90% CL).

T2K can also perform disappearance measurements of $\nu_\mu$, which will improve measurements of $\Delta m^2_{23}$ to a precision of 0.0001 eV$^2$ or so. Neutral current disappearance (in events tagged by $\pi^\circ$ production) will allow for a sensitive search of sterile neutrino production.

### 1.3.3.2 NOνA

The NOνA experiment with an upgraded NuMI off-axis neutrino beam [57] ($E_\nu \sim 2$ GeV and a $\nu_e$ contamination lower than 0.5%) and with a baseline of 810 km (12 km off-axis), has been proposed at FNAL with the aim to explore $\nu_\mu \to \nu_e$ oscillations with a sensitivity 10 times better than MINOS. The experimental schedule at the moment is not well defined. The NuMI target will receive a 120 GeV/c proton flux with an expected intensity of $6.5 \cdot 10^{20}$ pot/year ($2 \cdot 10^7$ s/year are considered available to NuMI operations while the other beams are normalized to $10^7$ s/year). The experiment will use a near and a far detector, both using liquid scintillator (TASD detectors). In a five-year $\nu_\mu$ run with a 25 kt active mass far detector, a $\sin^2 2\theta_{13}$ sensitivity slightly better than T2K, as well as a precise measurement of $|\Delta m_{23}^2|$ and $\sin^2 2\theta_{23}$, can be achieved. NOνA can also allow the mass hierarchy problem to be solved for a limited range of the $\delta_{CP}$ and $\text{sign}(\Delta m_{23}^2)$ parameters [57], this thanks to the longer baseline with respect to T2K.

As a second phase, proton intensity could be raised up to $10 \cdot 10^{20}$ pot/yr and a second detector placed into operation at a different off-axis angle at the second oscillation maximum [57].

### 1.3.3.3 *Reactor experiments*

Another approach to searching for non-vanishing $\theta_{13}$ is to look at $\bar{\nu}_e$ disappearance using nuclear reactors as neutrino sources.

Reactor experiments aim at improving the current knowledge on $\theta_{13}$ by observing the disappearance of $\bar{\nu}_e$ from nuclear reactors. The relevant oscillation probability is

$$P(\bar{\nu}_e \to \bar{\nu}_e) \simeq 1 - \sin^2 2\theta_{13} \sin^2 \left( \frac{\Delta m_{31}^2 L}{4E} \right) + \dots \qquad (1.8)$$

which does not depend on $\theta_{23}$ and the CP-phase $\delta_{CP}$. The dependence on $\Delta m_{21}^2$ and $\theta_{12}$ is negligible for the chosen baseline. Therefore this approach allows an unambiguous detection of $\theta_{13}$ free from correlations and degeneracies. As it is obvious from Eq. (1.8) the measurement requires a very precise control of the absolute flux. For that reason these experiments will employ a near and far detector. The direct comparison of the event rates in each detector will cancel many systematical errors and thus is essential in reaching the required low level of residual errors. Both detectors need some overburden to reduce the cosmic muon flux to an acceptable level.

The Double Chooz experiment, the follow-up to CHOOZ, will employ a far detector in the same location as the former CHOOZ detector as well as a near detector. Both detectors need some overburden to reduce the cosmic muon flux to an acceptable level. The advantage of Double Chooz is that it will use an existing cavern for the far detector, which puts it ahead of any other reactor experiment.

The sensitivity after five years of data taking will be $\sin^2 2\theta_{13} = 0.025$ at 90% CL [58], which could be achieved as early as 2012. It is conceivable to use a larger, second cavern to place a 200 t detector to further improve that bound down to $\sin^2 2\theta_{13} < 0.01$ [59].

The Daya Bay project in China [60] could reach a $\sin^2 2\theta_{13}$ sensitivity below 0.01 integrating 70 times the statistics of Double Chooz. This experiment will detect $\bar{\nu}_e$ produced by two pairs of reactor cores (Daya Bay and Ling Ao), separated by about 1.1 km. The complex generates 11.6 GW of thermal power; this will increase to 17.4 GW by early 2011 when a third pair of reactor cores (Ling Ao II) is put into operation.

It will consist of two near detector locations (each one with two 20 t detectors) and a far location consisting of four detectors of 20 t.

### 1.3.4 *Next generation conventional neutrino beams*

It is well-known today that T2K and NO$\nu$A, even if combined with a reactor experiment, will not be able to provide firm results ($3\sigma$ or better) about leptonic CP violation [61] or sign($\Delta m_{23}^2$) [62] whatever the value of $\theta_{13}$. A next generation of long-baseline neutrino experiments will be needed to address this very important search in physics. The rule of thumb in such experiments is that they should be at least one order of magnitude more sensitive than T2K or NO$\nu$A. As a result they need an increase of two orders of magnitude on neutrino statistics with a consequent important reduction of systematic errors.

#### 1.3.4.1 *Neutrino super beams*

To fulfill such a challenging improvement, conventional neutrino beams must be pushed to their ultimate limits (neutrino super beams) [29] and gigantic (megaton scale) neutrino detectors must be built.

In the following, a super beam is taken to be a conventional neutrino beam driven by a proton driver with a beam power in the range of 2-5 MW.

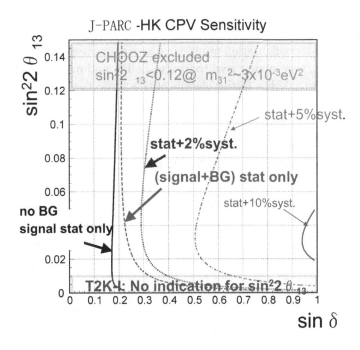

Fig. 1.7  LCPV sensitivity curve of T2HK. From [63].

### 1.3.4.2  *T2HK and T2KK*

Phase II of the T2K experiment, often called T2HK, foresees an increase of beam power up to the maximum feasible with the accelerator and target (4 MW beam power), antineutrino runs, and a very large, 520 kt, water Čerenkov, HyperKamiokande or HK, to be built close to SuperKamiokande.

Performances of such a setup have been computed in Ref. [63] (see Fig. 1.7). From this figure two main considerations can be taken:

(1) Systematic errors play a critical role in LCPV sensitivities: systematic errors bigger than 5% could potentially kill the experimental sensitivity. For a detailed discussion of systematic errors in T2HK and the possible role of ancillary experiments to reduce them, see [64].
(2) For $\theta_{13}$ values smaller than the T2K phase I sensitivity, T2HK has no sensitivity to LCPV. That is in case of a null result of T2K which would severely compromise the physics case of T2HK.

An evolution of T2HK is the T2KK project, where half of the HK detector would be installed in Japan, while the second half would be mounted in Korea, at a baseline of about 900 km, around the second oscillation maximum. This configuration would have worse sensitivities to $\theta_{13}$ with respect to T2HK, very similar sensitivities to LCPV and much better sensitivities to sign($\Delta m_{23}^2$). For further details of this project see [65].

### 1.3.4.3   *CNGS upgrades*

Sensitivities for an upgraded CNGS setup have been first computed in [66]. The experimental setup can be summarized as follows:

- A proton intensity of $4.1 \cdot 10^{20}$ pot/yr has been assumed, to be compared with the nominal CNGS intensity of $4.5 \cdot 10^{19}$ pot/yr [1].
- To be at the first maximum at a baseline of around 750 km, the beam optics has been re-designed to focus pions of about 10 GeV/c momentum, against the 35 GeV/c momentum of the nominal CNGS. Several different off-axis angles have been taken into account.
- A monolythic 100 kt liquid argon detector (ICARUS actually is composed by two tanks of 300 kt) should be installed at shallow depth at a baseline of about 800 km.

With this setup, in a ten-year $\nu_\mu + \overline{\nu}_\mu$ run a sensitivity of $\sin^2 2\theta_{13} \geq 1 \cdot 10^{-3}$ (3 $\sigma$, $\delta_{CP}$=0, sign($\Delta m_{23}^2$)=+1) could be reached.

A subsequent LoI, MODULAr [68], has been proposed by the ICARUS collaboration for a similar setup: $1.2\text{-}4.3 \cdot 10^{20}$ pot/yr and a modular 20 kt liquid argon detector to be installed at shallow depth at a baseline of 732 km, 10 km off-axis. This project could reach a sensitivity of $\sin^2 2\theta_{13} \geq 2.5 \cdot 10^{-3}$ (3$\sigma$) , $\delta_{CP}$=0, in five years of neutrino operation.

### 1.3.4.4   *CERN-SPL*

In the CERN-SPL super beam project [69, 70] the planned 4MW SPL (Superconducting Proton Linac) would deliver a 3.5 GeV/c H$^-$ beam on a Hg target to generate a neutrino beam with an average energy of $\sim 0.3$ GeV [2].

---

[1] According to the CERN report [67], the ultimate CNGS intensity, having changed the whole SPS injection chain, substituted the SPS RF cavities and upgraded the CNGS instrumentation, would be $1.3 \cdot 10^{19}$ pot/yr, or $2.45 \cdot 10^{19}$ pot/yr if all the fixed target programmes at CERN were canceled.

[2] At present SPL is foreseen as one of the elements of a new injection chain for the SPS, in view of the LHC luminosity upgrades [71]. In this context a power of 0.4 MW would be enough. Extensions to 4 MW could be driven by the needs of a neutrino super beam

The $\nu_e$ contamination from $K$ will be suppressed by threshold effects and the resulting $\nu_e/\nu_\mu$ ratio ($\sim 0.4\%$) will be known within 2% error. The use of a near and far detector (the latter at $L = 130$ km in the Fréjus area, see Section 1.3.4.6) will allow for both $\nu_\mu$-disappearance and $\nu_\mu \to \nu_e$ appearance studies. The physics potential of the SPL super beam (SPL-SB) with a water Čerenkov far detector with a fiducial mass of 440 kt, has been extensively studied [72–75].

The most updated sensitivity estimations for this setup have been published in Ref. [76] and are shown in Section 3.7.

### 1.3.4.5  *Wide-band super beam*

A wide-band beam (WBB) has been proposed, sited at BNL and serving a very long-baseline experiment [77]. In this proposal, the 28 GeV AGS would be upgraded to 1 MW and a neutrino beam with neutrino energies in the range 0 - 6 GeV could be sent to a Megaton water Čerenkov detector at the Homestake mine at a baseline of 2540 km.

Wide-band beams possess the advantages of a higher on-axis flux and a broad energy spectrum. The latter allows the first and second oscillation nodes in the disappearance channel to be observed, providing a strong tool to solve the degeneracy problem. On the other hand, experiments served by wide-band beams must determine the incident neutrino energy with good resolution and eliminate the background from the high energy tail of the spectrum.

Upgrades to the FNAL main injector after the end of the Tevatron programme are also under study and could provide a similar wide-band neutrino beam. The baseline in this case would be 1290 km.

The combination of channels and spectral information of a long-baseline wide-band beam experiment offers a promising means of solving parameter degeneracies. However, the very long-baseline decreases the event rate at the far detector and reduces the sensitivity of the experiment to $\theta_{13}$ and CP-violation; the sensitivity of the experiment to $\theta_{13}$ and $\delta$ is somewhat smaller than that of T2HK or the SPL.

The wide-band beam is a very interesting option to search for the sign($\Delta m_{23}^2$)value and for leptonic CP-violation, solving most of the degeneracies, if $\theta_{13}$ is large enough, i.e. $\sin^2 2\theta_{13} > 5 \times 10^{-3}$ ($\theta_{13} > 2°$).

---

or a proton driver for a neutrino factory and/or a proton driver for EURISOL.

### 1.3.4.6  *Water Čerenkov detectors and the MEMPHYS detector*

For small values of $\theta_{13}$, a very large data set is required for the sub-leading $\nu_\mu \to \nu_e$ oscillation to be observed. The water Čerenkov is an ideal detector for this task since it is possible to construct a detector of very large fiducial mass in which the target material is also the active medium. The Čerenkov light is collected by photo-detectors distributed over the surface of the detector; the cost of instrumenting the detector, therefore, scales with the surface area rather than the fiducial mass. Megaton-class water Čerenkov detectors are therefore ideal when charge identification is not required and have been chosen for T2HK, the SPL super beam, and the long-baseline wide-band beam experiment. Such a device could also be the ultimate tool for proton-decay searches and for the detection of atmospheric, solar, and supernovæneutrinos.

Charged leptons are identified through the detection of Čerenkov light in photo-multiplier tubes (PMTs) distributed around the vessel. The features of the Čerenkov rings can be exploited for particle identification. A muon scatters very little in crossing the detector, therefore, the associated Čerenkov ring has sharp edges. Conversely, an electron showers in the water, producing rings with "fuzzy" edges. The total measured light can be used to give an estimate of the lepton energy, while the time measurement provided by each PMT allows the lepton direction and the position of the neutrino interaction vertex to be determined. By combining all this information, it is possible to reconstruct the energy, the direction, and the flavor of the incoming neutrino. It is worth noting that the procedure discussed above is suitable only for quasi-elastic events ($\nu_l n \to l^- p$). Indeed, for non-quasi-elastic events more particles present in the final state are either below the Čerenkov threshold or are neutral, resulting in a poor measurement of the total event energy. Furthermore, the presence of more than one particle above threshold produces more than one ring, spoiling the particle identification capability of the detector.

The MEMPHYS (Megaton Mass Physics) detector [78] is a megaton-class water Čerenkov in straight extrapolation of the well-known and robust technique used for the Super-Kamiokande detector. It is designed to be located at Fréjus, 130 km from CERN, and is an alternative design of the UNO [79] and Hyper-Kamiokande [54] detectors and shares the same physics case in both the non-accelerator domain (nucleon decay, SuperNovae neutrino from burst event or from relic explosion, solar and atmospheric neutrinos) and the accelerator (super beam, beta beam) domain [185].

MEMPHYS may be built with current techniques near the present Modane Underground Laboratory as three- or four-shaft modular detector, with shafts of 250000 m$^3$ each, measuring 65 m in diameter, 65 m in height for the total water containment. Each of these shafts corresponds to about five times the present Super-Kamiokande cavity. For the current physical study, a fiducial volume of 440 kt which means three shafts and an Inner Detector (ID) 57 m in diameter and 57 m in height is assumed. A sketch of the possible layout of the laboratory is displayed in Fig. 1.8. Each ID may be equipped with photodetectors (PMT, HPD,..) with a surface coverage of at least 30%. The Fréjus site, 4800 m.w.e, offers a natural protection against cosmic rays by a factor of about 10$^6$.

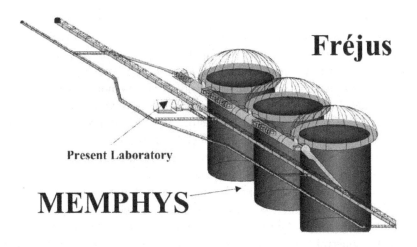

Fig. 1.8   Possible layout of the future Fréjus underground laboratory.

## 1.4   Why Look For New Concepts in Neutrino Beams?

The event collected in a neutrino oscillation detector, $N^{far}_{events}$, can be described by the following equation:

$$N^{far}_{events} = \left( \sigma_{\nu_e} \epsilon_{\nu_e} P_{\nu_\mu \nu_e} + \sigma^{NC}_{\nu_\mu} \eta_{NC} + \sigma^{CC}_{\nu_\mu} \eta_{CC} P_{\nu_\mu \nu_\mu} \right) \Phi_{\nu_\mu} + \sigma^{CC}_{\nu_e} \epsilon_{\nu_e} \Phi_{\nu_e} \quad (1.9)$$

where

- $\sigma_\nu(E_\nu)$ is the cross section of $\nu$
- $\epsilon_{\nu_e}(E_\nu)$ is the detection efficiency of electron neutrinos
- $P_{\nu_\mu \nu_e}(E_\nu)$, $P_{\nu_\mu \nu_\mu}(E_\nu)$ are oscillation probabilities
- $\eta_{NC(CC)}(E_\nu)$ is the detection efficiency of backgrounds from NC (CC) $\nu_\mu$interactions.
- $\Phi_{\nu_\mu}(E_\nu)$ is the $\nu_\mu$flux at the detector
- $\Phi_{\nu_e}(E_\nu)$ is the $\nu_e$ flux at the detector
- Cross section and fluxes are not known to better than 5%.

A close detector placed near enough the neutrino target to consider the oscillation probabilities negligible would measure

$$\mathrm{N}^{\mathrm{close}}_{\mathrm{events}} = \left( \sigma^{\mathrm{NC}}_{\nu_\mu} \eta'_{\mathrm{NC}} + \sigma^{\mathrm{CC}}_{\nu_\mu} \eta'_{\mathrm{CC}} \right) \Phi'_{\nu_\mu} + \sigma^{\mathrm{CC}}_{\nu_e} \epsilon_{\nu_e} \Phi'_{\nu_e} \qquad (1.10)$$

where the background efficiencies $\eta'_{NC(CC)}(E_\nu)$ are not necessarily the same as the far detector, and more important the neutrino fluxes $\Phi'_{\nu_\mu}(E_\nu)$ and $\Phi'_{\nu_e}(E_\nu)$ are not the same as the far detector. This latter phenomenon is due to the fact that the close detector is sensitive to the length of the decay tunnel, and for simple solid angle considerations is more sensitive to pions that decay late in the tunnel, the most energetic, while the far detector is insensitive to this geometrical factor. This effect produces differences between the close and far detector neutrino fluxes of up to 30%.

It is clear that a close detector cannot constrain all the single variables that contribute to the far detector signal rate, but first generation neutrino beams have demonstrated that it is still quite powerful in controlling systematic errors, especially if coupled with measurements of the pion production rates at the target (hadroproduction experiments) that can efficiently constrain the prediction of the neutrino flux both at the close and at the far detector sites. A quantitative discussion about the effects of the close detectors on neutrino super beam experiments can be found in [64].

With the close detector data the K2K experiment concluded the $\nu_\mu \rightarrow \nu_e$ search with an overall 30% systematic error [80]. This value could be improved by using the Harp hadroproduction data [44], as already done for the $\nu_\mu$ disappearance measurements of K2K.

The NOMAD experiment at CERN [81] was indeed able to predict the $\nu_e$ spectrum with a 5% systematic error [82]; this prediction was then successfully cross-checked with data [16]. The NOMAD experiment did not have a close detector but had the unprecedented capability of measuring with large statistics and good precision all four neutrino flavors ($\nu_\mu$, $\overline{\nu}_\mu$, $\nu_e$, $\overline{\nu}_e$) in a $\nu_\mu$ beam, a $\overline{\nu}_\mu$ beam and a neutrino beam without any horn focusing. The $\nu_\mu$, $\overline{\nu}_\mu$, $\overline{\nu}_e$ measurements in such conditions were basically enough to constrain the MC predictions of most secondary mesons in the neutrino beam line ($\pi^-$, $\pi^+$, $k^-$, $k^+$, $k^\circ$), reducing most of the systematic errors in the $\nu_e$ prediction.

The T2K experiment expects to push the systematics below 10%, thanks to an improved close detector system 280 m downstream the target. It will use detectors in a magnetic field to isolate the single neutrino flavors, and specialized detectors for the different backgrounds, in particular a specialized detector for the NC $\pi^\circ$ background. A second close detector station is under evaluation at a distance of 2 km from the target. At such a distance the neutrino fluxes are quite similar to the far detector fluxes and a water Čerenkov detector can be safely put into operation. This second station could be very powerful in further reducing the systematics.

The goal to constrain the systematics to less than 5% appears quite challenging in conventional neutrino beams. Many systematics that at first order cancel out in a close-far detector ratio contribute at the second order if the level of systematic errors is fixed at 5% or below, forcing the need to independently measure any single variable of Eq. (1.9).

Two main problems appear particularly critical from this perspective:

- The neutrino flux is necessary in any cross section and efficiency measurement. It requires a precise measurement of the neutrino hadroproduction at the target and a precise simulation of all the primary and secondary reinteractions in the target itself and in all the materials of the neutrino beam line. It should be noted that the best hadroproduction measurement published so far, the Harp measurement of the MiniBooNE beryllium target, concluded the analysis with a 4.9% integral systematic error and 9.8% differential systematic error [45].
- Conventional neutrino beams always produce at least four neutrino flavors ($\nu_\mu$, $\nu_e$, $\overline{\nu}_\mu$, $\overline{\nu}_e$), of which $\nu_e$ and $\overline{\nu}_e$ are an intrinsic irreducible background for any $\nu_\mu \to \nu_e$ search, while $\overline{\nu}_\mu(\nu_\mu)$ are a background for any CP search.

## 1.5 New Concepts on Neutrino Beams

The intrinsic limitations of conventional neutrino beams can be overcome if the neutrino parents are fully selected, collimated and accelerated to a given energy.

With this challenging approach several important improvements can be made to conventional neutrino beams:

- The neutrino fluxes would be simply derived from the knowledge of the number of parents circulating in the decay ring and from their Lorentz boost factor $\gamma$.
- The intrinsic neutrino backgrounds would be suppressed (in the case of beta beam) or reduced to wrong sign muons (golden channel in neutrino factories).

The technological problems derive from the fact that the parents need to be unstable particles, requiring a fast, efficient acceleration scheme.

This can be attempted within the muon lifetime, bringing to the neutrino factories [83], or within beta decaying ion lifetimes, bringing to the beta beam [85].

With this kind of beams (in the specific case of a beta beam) the number of events in the far detector would be given by

$$N_{\text{events}}^{\text{far}} = \left( \sigma_{\nu_\mu} \epsilon_{\nu_\mu} P_{\nu_e \nu_\mu} + \sigma_{\nu_e}^{\text{NC}} \eta_{\text{NC}} + \sigma_{\nu_e}^{\text{CC}} \eta_{\text{CC}} P_{\nu_e \nu_e} \right) \phi_{\nu_e} . \tag{1.11}$$

And in the close detector:

$$N_{\text{events}}^{\text{close}} = \left( \sigma_{\nu_e}^{\text{NC}} \eta_{\text{NC}}' + \sigma_{\nu_e}^{\text{CC}} \eta_{\text{CC}}' \right) \phi_{\nu_e} . \tag{1.12}$$

As a consequence:

- There is no need to disentangle NC from $\nu_\mu$ events at the close detector, since $\nu_\mu$ events are absent.
- Even if the close and far detector neutrino fluxes are different, they are fully predictable, so there is no need for a hadroproduction experiment with its associated errors.
- It should anyway be noted that the absence of $\nu_\mu$ events in the close detector means that there is no way to measure signal ($\nu_\mu$) cross sections in the close detector. This could represent the major source of systematic errors in a beta-beam experiment.

### 1.5.1 *Neutrino factories*

In a neutrino factory the neutrinos originate from the decay of muons which are created via pions by protons impinging on a target. The muons have to be captured, cooled, accelerated and put into a decay ring with straight sections pointing towards the detectors.

The muon decay is well understood. Consequently, the composition and spectral characteristics of the resulting neutrino beam can be calculated with high precision. Furthermore, the use of a decay ring will result in a well collimated neutrino beam. The intensity will scale with the power of the driver beam but it will also depend heavily on the efficiency of the capture and cooling schemes and the swiftness of the acceleration as muons are very short lived.

The IDS study [84] aims at $5 \times 10^{20}$ neutrinos along the straight sections of two independent storage rings of 25 GeV. This would require a driver beam intensity of some 4 MW at 5-10 GeV depending on the driver type. The target can either be a fast rotating solid target or a liquid metal target. Even metal power targets have been considered. Approximately a third of the driver beam power will be absorbed in the target with the remianing beam power being lost in a beam dump downstream of the target.

The pions emerging from the target are captured with a magnetic horn or a superconducting solenoid. The muons from the pion decay are cooled in an cooling channel in which ionization cooling is applied to reduce the beam emittance. The resulting muon beam is accelerated with recirculating linacs and a Fixed Field Alternating Gradient accelerator and finally stoored in two independent race track stoarge rings with straight sections of some 600 meter.

# Chapter 2

# Machine Aspects

## 2.1 Introduction

The challenges encountered in designing a beta-beam facility are very similar to those encountered when accelerating stable heavy ions. This is done at many accelerator labs in the world. In the Relativistic Heavy Ion Collider (RHIC) at Brookhaven National Laboratory (BNL) heavy ions are typically accelerated to 100 GeV per nucleon. At the Large Hadron Collider at European Organization for Nuclear Research (CERN) heavy ions will be accelerated to more than 2.5 TeV per nucleon (Fig. 2.1). The energy reached at BNL and CERN for heavy ions is more than sufficient for a beta beam. In fact, the first proposal made for a beta-beam facility [85] simply made use of the CERN injectors for the acceleration and only proposed the construction of an ion production part, a pre-accelerator and an additional decay ring to generate the neutrino beam. Ions are not more difficult to accelerate than any other particle.

The specific challenges for the ion accelerator designer are rather linked to the ion itself and its charge state. Only very light ions can be copiously produced and fully stripped of electrons at low energies. In fact, the proton is an example of a very light ion (hydrogen) which can be produced at high intensity, high density and fully stripped, making it the ideal particle for a hadron collider. At the other end of the spectrum, lead ions for LHC produced in the state-of-the-art Electron Cyclotron Resonance source will have more than 60 electrons left in atomic orbits at extraction from the ion source and the intensity and density will be much lower than for the proton beam.

The specific challenge for a beta-beam facility is the nature of the ions to be accelerated; the ions are not stable but they decay with a given lifetime

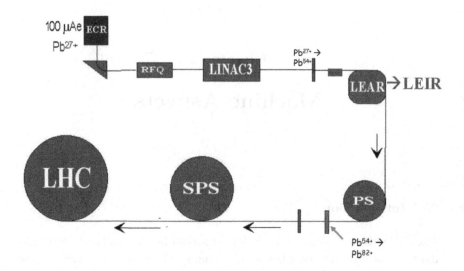

Fig. 2.1   A beta-beam facility at e.g. CERN would share many features (and much equipment) with the heavy ion programme at the Large Hadron Collider (LHC). Lead ions for LHC are produced in an Electron Cyclotron Resonance source and is, after accumulation in the Low Energy Ion ring, (LEIR) accelerated to more than 177 GeV/u per nucleon before being injected into the LHC. Illustration courtesy of the CERN ions for LHC project.

throughout the production and acceleration process.

## 2.2   A Possible Beta-Beam Facility

To simplify the discussion on the different parts of a beta-beam complex we will start with a description of a possible beta-beam facility as envisaged by a group of CERN machine physicists in 2002 [86]. This scenario was elaborated on just after the publication of the first paper by P. Zucchelli [85] on beta beams.

   In this design the ions are produced in a thick target using a proton beam of 1 GeV, extracted as neutral atoms and re-ionized and bunched in a high frequency ECR source. The first step of acceleration is a linear accelerator which brings the ions to a kinetic energy of 150 mega-electron volts (MeV) per nucleon. Subsequently, this beam is injected into a Rapid Cycling Synchrotron (RCS) in which the energy is increased to 500 MeV per nucleon. After this first step the beam enters the existing CERN accelerator complex and is accelerated in the Proton Synchrotron (PS) to its maximum energy, transferred to the Super Proton Synchrotron (SPS) and finally ejected to a decay ring. The last step is done using a scheme which permits the "new" ions to be merged with the ions already circulating in the decay ring so that ions which still can decay (and create neutrinos) are not wasted and ejected from the decay ring too early. The details of each step in this scheme will be discussed in detail below. To help the reader follow the different steps an overview of the facility is shown in Fig. 2.2.

To follow the different sections of this chapter it is only important to retain the following: i) that a sequence of accelerators are used to accelerate bunches of ions ii) that the ions are produced continuously in a target and bunched and iii) that the ions are accumulated in a few short bunches in a decay ring with long straight sections to generate a pulsed neutrino beam.

The essence of the CERN proposal in 2002 was the re-use of existing heavy ion accelerators to reduce cost and gain time for the construction of a beta-beam facility. With a similar intention a study was undertaken in the US in 2004 investigating the possible use of existing accelerators at Brookhaven National Laboratories and Fermi National Laboratories [87].

## 2.3   The Beta Beam Isotopes

### 2.3.1   *Which isotope to use*

The ideal beta beam isotope should be sufficiently long-lived not to decay during the acceleration process and sufficiently short-lived to decay in the decay ring before it is lost due to other processes.

The use of very long-lived nuclei will also result in a very high total current in the decay ring which in itself might be a limiting factor due to the inter-ion collisions within each bunch (intrabeam scattering) and the problem of controlling such a number of charges (space charge) in the ring. Furthermore, the ideal isotope should be easy to produce in large quantities and it should not decay to any long-lived daughter products which could contaminate the low energy part of the accelerator chain and make maintenance work on the machine components difficult. At higher energies the decay products will break down and form hadron cascades as soon as they collide with the surrounding materials. Finally, for all ions except for the lightest elements such as helium, lithium and boron, the ions will be difficult to strip of all surrounding electrons at low energy. This will

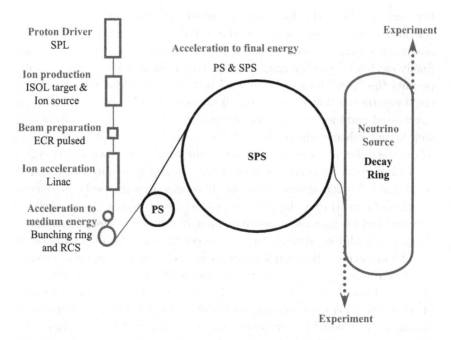

Fig. 2.2 An overview of a possible beta-beam facility at CERN as discussed in [86]. Note that the existing accelerators PS and SPS were proposed as part of the injector chain, representing a major saving for the proposal.

represent a significant intensity loss as it is hard to force more than 20% of the ions of a particular type to be in the same charge state.

The first study of a beta beam at CERN established a list of suitable isotopes taking all the above considerations into account; this list is reprinted in Table 2.1 ($\beta^-$ emitters) and Table 2.2 ($\beta^+$ emitters). For this first study $^6$He and $^{18}$Ne were selected; they have half-lives at rest in the order of a second which corresponds to the typical cycling time of low energy accelerators, they are easy to produce, are in gas phase at room temperature and have no "dangerous" daughter products.

### 2.3.2 *Isotope production*

Radioactive ions must be produced continuously as there is no way to stop them from decaying. There are two different methods used in modern Nuclear Physics for on-line production of exotic radioactive ions: the In-Flight (IF) method and the Isotope Separation On-Line (ISOL) method [88]. With

Table 2.1   Some possible isotopes which are $\beta^-$ emitters, from [86].

| Isotope | A/Z | $T_{1/2}$ (s) | $Q_\beta$ g.s. to g.s. (MeV) | $Q_\beta$ effective (MeV) | $E_\beta$ average (MeV) | $E\nu$ average (MeV) |
|---|---|---|---|---|---|---|
| $^6$He | 3.0 | 0.80 | 3.5 | 3.5 | 1.57 | 1.94 |
| $^8$He | 4.0 | 0.11 | 10.7 | 9.1 | 4.35 | 4.80 |
| $^8$Li | 2.7 | 0.83 | 16.0 | 13.0 | 6.24 | 6.72 |
| $^9$Li | 3.0 | 0.17 | 13.6 | 11.9 | 5.73 | 6.20 |
| $^{11}$Be | 2.8 | 13.8 | 11.5 | 9.8 | 4.65 | 5.11 |
| $^{15}$C | 2.5 | 2.44 | 9.8 | 6.4 | 2.87 | 3.55 |
| $^{16}$C | 2.7 | 0.74 | 8.0 | 4.5 | 2.05 | 2.46 |
| $^{16}$N | 2.3 | 7.13 | 10.4 | 5.9 | 4.59 | 1.33 |
| $^{17}$N | 2.4 | 4.17 | 8.7 | 3.8 | 1.71 | 2.10 |
| $^{18}$N | 2.6 | 0.64 | 13.9 | 8.0 | 5.33 | 2.67 |
| $^{23}$Ne | 2.3 | 37.2 | 4.4 | 4.2 | 1.90 | 2.31 |
| $^{25}$Ne | 2.5 | 0.60 | 7.3 | 6.9 | 3.18 | 3.73 |
| $^{25}$Na | 2.3 | 59.1 | 3.8 | 3.4 | 1.51 | 1.90 |
| $^{26}$Na | 2.4 | 1.07 | 9.3 | 7.2 | 3.34 | 3.81 |

Table 2.2   Some possible isotopes which are $\beta^+$ emitters, from [86].

| Isotope | A/Z | $T_{1/2}$ (s) | $Q_\beta$ g.s. to g.s. (MeV) | $Q_\beta$ effective (MeV) | $E_\beta$ average (MeV) | $E\nu$ average (MeV) |
|---|---|---|---|---|---|---|
| $^8$B | 1.6 | 0.77 | 17.0 | 13.9 | 6.55 | 7.37 |
| $^{10}$C | 1.7 | 19.3 | 2.6 | 1.9 | 0.81 | 1.08 |
| $^{14}$O | 1.8 | 70.6 | 4.1 | 1.8 | 0.78 | 1.05 |
| $^{15}$O | 1.9 | 122 | 1.7 | 1.7 | 0.74 | 1.00 |
| $^{18}$Ne | 1.8 | 1.67 | 3.3 | 3.0 | 1.50 | 1.52 |
| $^{19}$Ne | 1.9 | 17.3 | 2.2 | 2.2 | 0.96 | 1.25 |
| $^{21}$Na | 1.9 | 22.4 | 2.5 | 2.5 | 1.10 | 1.41 |
| $^{33}$Ar | 1.8 | 0.17 | 10.6 | 8.2 | 3.97 | 4.19 |
| $^{24}$Ar | 1.9 | 0.84 | 5.0 | 5.0 | 2.29 | 2.67 |
| $^{35}$Ar | 1.9 | 1.77 | 4.9 | 4.9 | 2.27 | 2.65 |
| $^{37}$K | 1.9 | 1.22 | 5.1 | 5.1 | 2.35 | 2.72 |
| $^{80}$Rb | 2.2 | 34 | 4.7 | 4.5 | 2.04 | 2.48 |

the In-Flight method the ions are produced through fragmentation of a heavy stable beam in a thin target. The produced ions are separated after the target in an electro-static mass and velocity filter resulting in a rather large emittance but isotopically pure beam.

The ISOL method (see Fig. 2.3) uses a thick target in which a beam of particles is almost stopped. At higher energies this will result in

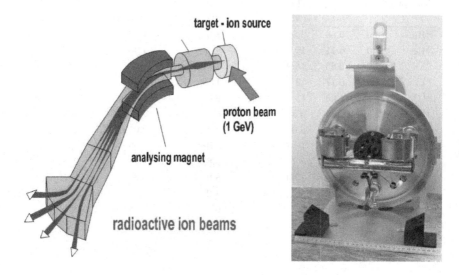

**target - ion source**

**proton beam (1 GeV)**

**analysing magnet**

**radioactive ion beams**

Fig. 2.3   At most operation ISOL facilities, a thick and hot target is used (right). The driver beam induces nuclear reactions throughout a large volume of the target and any isotope with a "boiling point" below the temperature of the assembly will diffuse out of the target matrix. A large fraction of the produced activity will reach the ISOL source and will after ionization be extracted into a magnetic separator (left). Illustrations courtesy of the ISOLDE collaboration and the CERN photo laboratory.

fragmentation, spallation and fission of the target material (and the impinging particles) while at very low energies the new isotopes are formed by the merging of the target nucleus and the incident nucleus. The thick target represents a high integrated cross section and the ISOL method will typically produce much higher intensity beams than the In-Flight method. However, the radioactivity produced in the thick target must be diffused and effused out of the target before it can be re-ionized and separated for further use. To enhance diffusion and effusion out of the target it must be heated. For high intensity facilities the heating caused by the beam is more than sufficient and the challenge for the target designer is to get rid of the excess heat so that the target is not destroyed. For the shortest-lived elements with high boiling points the target has to be kept at a temperature just below the melting point of the target material itself.

### 2.3.3 *The ISOL method*

For the beta beam, the intensity is the main issue so the ISOL method – or one of the variants of the ISOL method discussed in the next subsections – is the natural choice for the production part. The ISOL method encompasses several rather different techniques, all with some advantages and some drawbacks. A major issue is the heat deposited and damage caused by the incident charged particles.

A particularly attractive ISOL technique which partly resolves this issue is the converter technique in which the incident beam does not hit the target itself but a very robust primary target in which huge amounts of neutrons and protons are produced through spallation and evaporation processes. The protons released from the primary target are charged and are highly likely to be absorbed by the target itself while the neutrons are unlikely to be stopped. If we place a secondary target of fissile material in the surrounding neutron flux we will induce fission but without the damage which the charged protons would have caused. Unlike the secondary target the primary target can be cooled and made of a material highly resistant to ionizing radiation so as to take a large current of protons. A converter target for the production of $^6$He has been studied [89, 90] using the reaction $n + {}^9Be = {}^6He + {}^4He$ which has a large cross section of some 100 mbarn. To the left in Fig. 2.4 the conceptual design of the beryllium-oxide target and Tungsten converter system from [89] is shown. The beryllium-oxide is either in fiber form or pellets which will permit the Helium atoms to diffuse out of the target. The beam used to generate the neutrons from the converter is a 1 GeV proton beam and the resulting relative neutron flux is shown to the right in Fig. 2.4. The production of $^{18}$Ne with a thick ISOL target has also been studied [86] but for this proton-rich unstable nucleus, it is not possible to use the converter technology. The cross section, for example, of a 1 GeV proton beam impinging on a magnesium oxide target is at least a factor of ten smaller than the cross section for $^6$He production via neutrons on beryllium oxide. The fact that the target will be directly irradiated by a high intensity proton beam will result in radiation damage in the target setting a limit for the achievable yield compared to the converter technology. A possible way forward is to use multiple target units and merge the neutral atomic beams before ionization (see Fig. 2.5). This concept was tested at ISOLDE at CERN in 2006 [91] and for a configuration with two transfer lines a high efficiency for the merging into a plasma ion source was achieved.

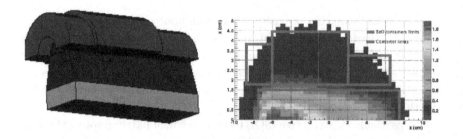

Fig. 2.4 To the left: a converter target for the production of $^6$He with the cooled converter (lower part) made of tungsten and the target made of beryllium oxide (covering upper part of converter) [89]. To the right: the relative neutron flux from the converter which is hit by a 1 GeV proton beam [89].

Fig. 2.5 For production of Ne through the spallation of e.g. MgO, the use of several targets at 1 – 2 GeV has been envisaged. In the picture the primary proton beam would be wobbled over the multiple target units. Illustration courtesy of Stefano Marzari at CERN.

### 2.3.4 *Direct production*

The simplest way to create nuclei is to accelerate one nucleus and merge it with another nucleus in a target at an energy high enough to overcome the Coulomb barrier, but low enough not to destroy the newly formed nuclei through spallation or fission. Nuclei formed in this energy interval are referred to as compound nuclei; cross section for this process is usually large and can often be measured in 10–100 mbarns. The main limitation is that it is hard to form any nucleus far away from stability as the starting point usually is two stable nuclei with roughly equal numbers of protons and neutrons. Consequently, the new nucleus will also have roughly the same number of neutrons as protons which will position it somewhere close to the other stable nuclei. However, for beta beams the favored isotopes are close to stability so this production method is a possibility.

The neon isotope discussed earlier in this section can be formed in the reaction

$$^{16}O +^3 He =^{18} Ne + n$$

or as a nuclear physicist would write

$$^{16}O(^3He, n)^{18}Ne.$$

The process has been studied in detail [92] and is usually referred to as direct production. The cross sections are indeed large for the reaction above but to produce a sufficient number of $^{18}Ne$ isotopes for a beta-beam facility using a MgO solid target, 120 mAmps of primary $^3$He beam at some 13 MeV of total energy is required. This is far beyond what has been done so far and would require the development of a new concept for the low energy beam dump.

For the high intensity beam required for a beta-beam production facility the target would be destroyed if it also had to cope with the full beam heating from the stopping $^3$He ions. To overcome this problem the target is made sufficiently thick to maximize the production but still thin enough to let through the remaining low energy ion beam so that it can be dumped in e.g. a liquid metal cooled beam dump. If the operating direct production facility in Louvain-la-Neuve in Belgium at the Cyclotron laboratory is taken as a reference, the target of e.g. MgO would have to be 60 cm in diameter to keep the power density at the level of the one used today. The proton beam would have to be de-focused an wobbled over the target but providing that a sufficiently intense $^3$He beam can be produced this scenario seems feasible.

In Fig. 2.6 the total cross section for the reaction above is plotted as a function of the projectile energy. Note that below the Coulomb barrier at a few MeV the cross section is vanishing which means that the low energy tail of the ions can be dumped in a passive beam dump without any loss of intensity.

### 2.3.5 *Production ring*

In direct production facilities the part of the beam which does not produce a new isotope through nuclear reaction is simply lost in a beam dump. To

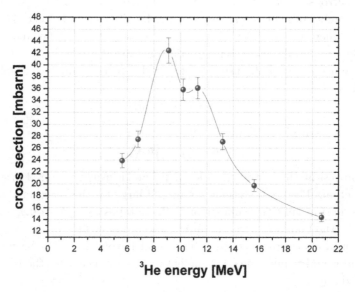

Fig. 2.6   The measured cross section for producing $^{18}$Ne with a $^{3}$He beam in the reaction $^{16}O(^{3}He,n)^{18}Ne$ from [92].

avoid this "waste" of useful ions they could be re-circulated, re-accelerated and sent through the target again. If the target is made sufficiently thin the ions can be made to pass at the optimum energy for the desired reaction channel each time ensuring that the majority of ions that react will produce a "useful" ion. The limiting factor seems, at a first glance, to be the angular straggling which eventually would make the re-circulated beam too "large" to handle. However, it was recently shown [93, 94] that the combination of energy loss in transverse directions in the target with re-acceleration to compensate for the lost energy will result in a net beam cooling. In [93] a wedge-shaped gas target is used in a dispersive region of the ring which adds longitudinal cooling as particles with higher energy can be made to pass through a thicker part of the target compared to those with lower energy. Furthermore, the use of a gas target makes it possible to handle a large amount of beam power. The produced ions are collected with a second target consisting of e.g. tantalum foils contained in a box with a hole through the center in which the circulating beam can pass the target without interacting with the foils (see top part of Fig. 2.7). The produced ions will be thermalized and neutralized in the foil, diffuse to the foil boundary as a neutral gas and through random walk in an effusion process find the exit of the box where they are re-ionized and extracted for bunching and further acceleration. The proposed reaction channels are $^7Li(d,p)^8Li$ and $^6Li(^3He,n)^8B$, both assuming a gaseous target and inverse kinematics (projectile lighter than target).

The isotopes $^8Li$ and $^8B$ emit higher energy neutrinos than $^6He$ and $^{18}Ne$ and could be used for a beta-beam facility with a longer baseline than the the proposed EURISOL beta-beam facility. In [94] a Fixed Field Alternating Gradient (FFAG) accelerator with large longitudinal acceptance is used to manage the beam without any longitudinal cooling. For both machines the beam is injected partially stripped and the energy of the circulating ions is kept high enough to ensure that all of them emerge fully stripped after the target. The production of $^8B$ and $^8Li$ with $^3He$ and deuterium as projectiles and a liquid lithium target of enriched $^6Li$ or $^7Li$ has been proposed in [95] in which also a full six-dimensional analysis of the cooling process is presented. A solid target would quickly overheat and burn and be difficult to make thin enough to avoid an important re-absorption of the produced ions. The thin liquid Li film could be produced with a high pressure jet directed at an angle towards a flat deflector as proposed and studied by [96]. The larger difference in magnetic rigidity between the projectile and the produced ion in this kinematic could permit beam collection off-axes using e.g. a Wien filter after the target to separate the secondary ions from the circulating primary ions (see Figure 2.8 and bottom part of Figure 2.7). Note that if a Wien filter is used the primary beam may have to be brought back to the nominal closed orbit with a "reversed" Wien filter further downstream. The physical separation of beam and produced ions will also reduce the background of beam particles deviated to large angles – through simple (single) Rutherford scattering in the target – in the collection device and it increase the total efficiency of the collection as there is no need for a "hole" in the collector.

### 2.3.6 *Production rates*

The achievable production rate in a thick (ISOL) target can be estimated using known cross sections and known diffusion and effusion parameters. However, such estimates have large uncertainties without experimental verification of the input data. Still, with the objective of giving a reasonable range for the annual rate of (anti)neutrinos at a beta-beam facility, the order of magnitude which is believed within reach with the different methods

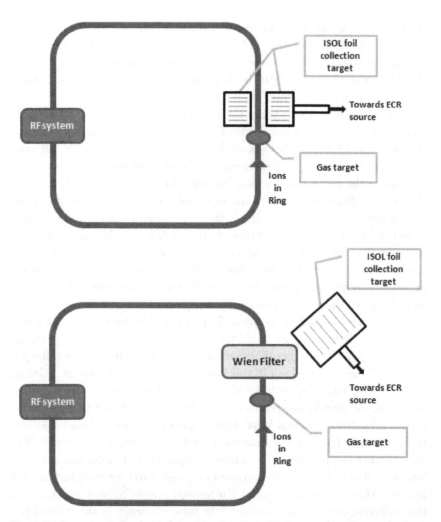

Fig. 2.7   Conceptual sketch of the production ring. In the top part the ring is shown with a collection device as proposed in [93]. In the bottom part the ring is shown with a collection device off-orbit using a Wien filter to deviate the ions of interest towards the device and to suppress some of the single scattered beam particles.

discussed in this section is given in Table 2.3. Furthermore, we are giving limits for the production rate in the target and it should not be forgotten that the extraction efficiency will vary considerably from a few percent up to 90% for the number of (charged) ions extracted.

Only detailed studies of all parts of the target and ion source system can give reliable numbers for the extraction efficiency.

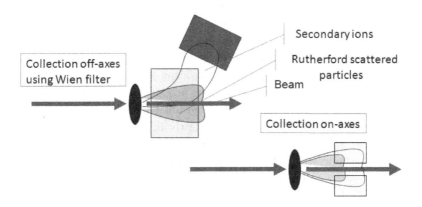

Fig. 2.8 Schematic representation of two different collection scenarios for the production ring concept. In the bottom-right part of the figure, the collection scenario proposed in [93] with a box with a hole through the middle for the circulating beam. The box contains foils in which the produced ions are stopped and thermalized. The ions will diffuse out of the hot foils and they will be collected and re-ionized for further ionization as in an ISOL target. The cascade of Rutherford scattered beam particles from the primary target will also be collected in the device. In the top-left part of the figure the collection is down off-axes with a Wien filter creating a transverse separation between the circulating beam and the produced ions.

Table 2.3 Estimates made by the authors for the production rate per second in the target of a few isotopes of interest for beta beams. Note that the references are to the methods rather than to the production limits.

| Isotope | Method | Rate within reach ions/second | Reference |
|---------|--------|-------------------------------|-----------|
| $^{18}$Ne | ISOL at 1 GeV and 200 kW | $< 8 \times 10^{11}$ | [86] |
| $^{6}$He | ISOL converter at 1 GeV and 200 kW | $< 5 \times 10^{13}$ | [86] |
| $^{18}$Ne | Direct production through $^{16}$O($^{3}$He,n)$^{18}$Ne | $< 1 \times 10^{13}$ | [92] |
| $^{6}$He | ISOL converter at 40 MeV Deuterons and 80 kW | $< 6 \times 10^{13}$ | [90] |
| $^{8}$Li | Production ring through $^{7}$Li(d,p)$^{8}$Li | $< 1 \times 10^{14}$ | [93] |

## 2.4 Ion Transfer, Ionization and Bunching

The ions produced in a thick ISOL target has to be collected and ionized before they can be accelerated. As explained in the previous section the target is heated to a temperature above the "boiling point" of the element to be extracted and is collected as a neutral gas which will effuse out of the target container, through a transfer line and into an ion source.

Experiments with noble gases have demonstrated that the velocity of the neutral gas is in the order of 1 m/s so the distance between the target and ion source should be kept as short as possible to avoid important decay losses before ionization. The ideal ionization and bunching system for beta beams should have very high efficiency, produce ions in a single high charge state and bunch the beam, and it should be simple and highly resistant to radiation. The bunching is necessary for the injection into circular machines for further acceleration, a process which will be discussed in some detail later. The bunching of ions for stable ion accelerators is usually done with an ECR source, EBIS source or a Duoplasmatron source. These systems are usually fed from a gas-bottle for gaseous elements such as Helium and Neon but they operate at a relatively low efficiency (a few percent). For stable ions this is not a problem as the gas flow can be easily increased to compensate for the shortfall but for the beta beam this would represent an unacceptable loss of efficiency. In addition, the individual pulses of ions required for the beta beam represent a large number of charges which cannot be handled with e.g. an EBIS source.

It is beyond the scope of this text to describe the details of each ion source and buncher type. For our purpose it is sufficient to know that an ECR source operates with a plasma generated with a radiofrequency source and confined with a magnetic field. For the bunching of the beam the ECR source is normally operated in a mode which favors capture and confinement of the gas of interest and the ions are ejected from the source by simply turning off the radiofrequency source. This effect is called the afterglow and is used at CERN in the ion source and buncher for the LHC ion programme. An alternative approach is to fill the ion source with neutral gas and contrary to the afterglow mode, the ions are now ionized and ejected when the radiofrequency source is turned on, and so called the pre-glow mode [97]. The frequency has to be a lot higher for the number of ions required for a beta beam and the magnetic confinement is complex so this concept still requires major R&D before it can be considered operational. For the afterglow operation the ions must be ionized to a low charge state before injection into the ECR to reach good capture efficiency (neutral gas injection into the afterglow operation has been tested but with limited success). This can be done with a high efficiency monoECR source which would also simplify the transport of the ions from the target area to the buncher.

It is at this stage worth introducing the concept of emittance (with apologies to readers who already are familiar with the basics of particle beams). The transverse emittance of a beam is set by the transverse beam size and the angular divergence of the beam. The transverse emittance is often written as $\epsilon_t$ and can be visualized as a surface in a two-dimensional plot in which the beam size is plotted versa the beam divergence. For ions at energies relevant for beta beams the surface spanned by these two parameters remains constant and cannot be changed by any conservative force e.g. focusing. In fact, the beam simply behaves like any incompressible liquid and the starting point for any more scientific discussion of emittances is taken from Liouville's theorem. The emittance will become smaller during acceleration – an effect which in accelerator physics is called adiabatic damping – but the normalized emittance, expressed as $\epsilon_t^* = (\beta\gamma\epsilon_t)$ in which $\beta$ and $\gamma$ are the Lorentz factors, will remain invariant even during acceleration. To change the emittance, some form of "beam cooling" has to be used and we will discuss that briefly in the section on stacking. For the beta beam, the initial emittance of the beam will be set by the ion source and buncher.

Finally, it should also be mentioned that a charge exchange scheme [98] was proposed at an early stage for the bunching of the beam. For this type of operation the partly stripped ions are injected into a stripping foil positioned in the closed orbit of a storage ring. The stripping will force the ions onto the closed orbit of the ring and with the help of a radio-frequency cavity the DC beam can eventually be bunched. However, there are severe limits on the number of foil passages the circulating beam can support due to angular straggling and the number of charges which can be handled for a given aperture, which makes this option less interesting for the beta beam.

## 2.5  Acceleration

The non-accelerator specialist is most likely to have heard about accelerators in the context of the many synchrotron light sources in which electrons are accelerated or in the context of medical machines for cancer treatment and isotope production using protons. Both particle types were accelerated early on in the history of accelerator technology and it is probably fair to say that the basic equations guiding accelerator design were initially derived with protons and electrons in mind.

The two types of accelerators needed for protons and electrons respectively differ in important aspects due to the difference in mass between the particles (the difference in charge only reverses the polarity of the system) and for the beta beam it is only worth retaining the proton accelerator as a reference. In this text we will only deal with the main issues for ion (and proton) acceleration and for a more detailed discussion we refer to *e.g.* [99].

Accelerating a heavy ion is more challenging than accelerating a proton as the charge per mass unit $(Q/A)$ is smaller and so-called charge-exchange reactions are more severe. The first will make acceleration and transverse and longitudinal focusing more demanding while the latter will increase the losses during acceleration and storage. For the beta beam the losses due to radioactive decay during acceleration and storage is an additional complication.

### 2.5.1 *Linear accelerators*

The simplest and fastest form of accelerator, but also the most space demanding and costly, is the linear accelerator (linac). The possibility of the fast acceleration of a (semi-)continuous beam in which the ions make a single passage through each accelerating cavity, thereby avoiding some collective phenomena which are particularly severe at low energy, makes the linac the ideal first stage of a beta-beam facility. Such a radioactive ion linac, albeit at lower energy and lower intensity, is already in use at e.g. TRIUMF's ISOL facility ISAC-II (see Fig. 2.9). In principle, a linac can also be used to accelerate more than one charge state [100] but this will add to the cost and be more complicated to operate. In a multi-charge-state accelerating linac the differently charged ions accelerated will be brought into the single maximum charge state available using a simple electron stripping stage at a sufficiently high energy. For the beta beam, light ions such as $^6$He and $^8$Li will predominantly emerge fully stripped from the bunching stage after the target and ion source while heavier ions such as $^{18}$Ne will exhibit a charge-state distribution. In the studies done so far the multiple charge-state acceleration technique has not been explored. The cost of a linac is to first order proportional to its length. This is determined by the requested top energy, the available integral voltage gradient along the linac and the $A/Q$ of the accelerated particles. A recent study of high intensity ion linacs [101] has shown that a suitable compromise between cost and performance using SC cavities technology results in a top energy of beta-beam ions of 100–200 MeV per nucleon after the linac.

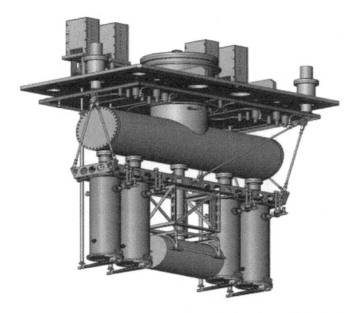

Fig. 2.9    The ISAC-II facility at TRIUMF consists of a linac in which the higher energy stage acceleration is done with Super Conducting Quarter wave cavities. The cavities are grouped four by four in cryostats keeping them at liquid helium temperature. The image shows the cavities (and a solenoid) before being inserted into the cryostat. Illustration courtesy of Robert Laxdal at TRIUMF.

### 2.5.2    *Some basics about synchrotrons*

The synchrotron is basically a linac in which the particles are bent with dipole magnets into a closed orbit and return to the same radiofrequency cavities over and over again for acceleration. The injector – which could be a linac – fills the machine with particles, but as soon as acceleration has started and the field in the dipoles is increased to keep the particles on a closed orbit, no new beam can be injected. This leaves the injector of the synchrotron hibernating during the cycling of the synchrotron and gives the accelerated beam a highly bunched time structure set by the length of the acceleration cycle. This renders the synchrotron much less efficient than a linac for the production of high power beams. Furthermore, the fact that the accelerated beam sees the same elements over and over again makes it more sensitive for different electromagnetic field imperfections in the machine resulting from e.g. element misalignments.

In a synchrotron, the beam performs oscillations in time (and distance) about a closed orbit near the center of the beam pipe. These oscillations will be determined by the focusing (and defocusing) elements in the ring, and the setting of all the magnetic elements, which is usually called the "lattice" of the machine. The lattice determines the shape of the beam envelope around the ring, and determines the number of oscillations performed by the beam as it moves along the closed orbit. This number, the "tune", is normally given as $Q$, with a fractional part $q$. If the field imperfections seen by the beam as it oscillates around the ring appear resonant with the beam oscillations, the beam will be lost (see Fig. 2.10).

### 2.5.2.1 *Space charge*

The electrostatic and magnetic forces from the many particles which make up each individual bunch – the "space charge" – will add (or subtract) to the focusing in both a coherent and incoherent way. The result of the incoherent part is that there will be a certain spread in the number of oscillations, $Q$, between individual particles in each beam bunch and this will make it difficult to avoid instabilities. This phenomenon of incoherent tune spread was first described in [102] which also derived a formula to estimate the space charge (de)focusing of the beam. The incoherent spread is expressed as $\delta q$ which is the spread of the number of oscillations, $Q$, performed by the beam as it moves along the closed orbit. So far we have discussed the instabilities in completely general terms but to understand how we can manage some of the higher order effects it is worth considering how the instabilities are induced. If the number of oscillations $Q$ is (almost) an integer – so-called dipolar imperfections – it is "dangerous" as the beam will return to the point of the imperfection with the same phase (see Fig. 2.10). The oscillation amplitude will increase turn after turn until the particle is lost. For a half-integer number of oscillations the particles will return to the quadrupolar imperfections with the same phase every second turn which will slow down the amplitude increase compared to the dipolar effect. For higher order effects the time the instabilities take to grow will be even longer. The experience from proton machines is that while any rational number of $Q$ in principle will cause instabilities, the longer time constant and the fact that it is possible to compensate for higher order effects makes a tune spread ($\delta q$) of 0.25 manageable. The tune shift can be expressed as

$$\delta q = -\frac{3Z^2 r R N \pi}{4A\beta^2\gamma^3 L_b \epsilon} \tag{2.1}$$

where $R$ is the average radius of the machine, $r$ is the proton radius, $L$ is the individual bunch length, $N$ the number of ions per bunch, $Z$ the charge of the ion, $A$ the mass of it, $\epsilon$ the transverse rms emittance of the bunch and $\beta$ and $\gamma$ the Lorentz factors. The other constants are given by the bunch shape (taken as pure Gaussian in this example) and the machine vacuum chamber geometry. The important thing to note is that for a given machine of a given circumference we have no real possibility of changing the $\delta q$ except by distributing the charges on an evenly spread number of transversely large and in time flat bunches. Big gaps between the bunches, very short and sharp bunch shapes and/or a small transverse beam emittance will all result in a large tune shift. As we want to have as many ions as possible to achieve a high intensity neutrino beam, the space charge limitations will be very important and will ultimately determine the absolute limit of any beta-beam facility in which we use circular machines.

There are evidently many other effects which might limit the number of ions in the accelerator before the space charge limit is reached, such as beam loading of the accelerating cavities. Similarly, the beam impedance of the ensemble of vacuum chambers might induce other types of beam instabilities, such as multi-bunch instabilities. However, one can argue that the space charge effects are a result of the fundamental design (e.g the circumference, energy range etc.) parameters and cannot be changed for any existing accelerator.

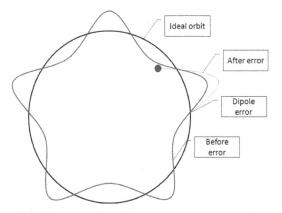

Fig. 2.10    The figure illustrates how a single dipole error increases the amplitude of the betatron oscillation for a particle with a $Q$ of 5. The amplitude will increase with each turn and eventually the particle will be "kicked out" of the machine.

### 2.5.2.2   *Injection and acceleration in synchrotrons*

The semi-continuous beam from the linac is injected into the first synchrotron over a large number of turns. The combination of the chosen
working point ($Q$) and a deliberate shift of the central beam orbit with
dipolar magnets makes it possible to inject over 50–100 turns with only
moderate losses ($< 30\%$) [103](see Fig. 2.11). The radiofrequency cavities
will be slowly ramped up once the injection process is completed forcing
the beam into the potential walls (or RF buckets as accelerator physicists
names them) and enabling the acceleration process to start. The speed of
the acceleration process is determined by the available RF voltage and the
ramp rate of the magnetic field in the magnets. It is possible to build synchrotrons which operate with up to 50 Hz repetition rates for a change in
the bending dipoles magnetic field with a maximum factor of 10. However,
most of the larger synchrotrons, e.g. the PS and SPS at CERN, operate
with much lower repetition with cycle times of several seconds.

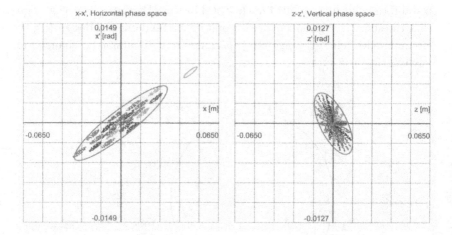

Fig. 2.11   The multi-turn injection from the linac into the Rapid Cycling Synchrotron
(RCS) at the CERN beta-beam facility has been simulated [103]. The transverse phase
space of the RCS can be efficiently filled with up to 40 turns of semi-continuous beam
from the linac using a combination of dipoles which effect the closed orbit in the ring
and septa magnets which steer the injected beam.

### 2.5.2.3 *Beam losses*

The manipulation of the ion beam between the different accelerators for e.g. stripping and injection will induce losses. Furthermore, the collision with rest gas in the accelerator itself and the decay of radioactive ions will add to these losses. The lost ions can both induce radioactivity and cause vacuum degradation.

To induce radioactivity the energy must be high enough for nuclear reaction to occur (some MeV/u) while a vacuum degradation can be caused at any energy through the secondary release of stable ions from the vacuum chamber walls as the particles hit the inside of the chamber. Both effects are best studied with Monte Carlo codes [104, 105]. The loss pattern along the circumference of any synchrotron will be determined by the size of the vacuum chamber and the lattice. The ions which change charge due to gas collisions and decay products will continue on new equilibrium orbits determined by the new mass-to-charge ratio, and as the machine is not tuned to transport these particles they are highly likely to be lost e.g. at the first obstacle in the new equilibrium orbit. Here there is an important difference between using existing synchrotrons (built for e.g. feeding hadron colliders) or using a machine specifically designed for the acceleration of Radioactive Ion Beams (RIB); in an existing machine the losses have a tendency of being equally distributed over the circumference while for a machine designed for RIBs the losses can be concentrated to certain sections in which collimators and absorbers (see Fig. 2.12) can help reduce the negative effects of the losses [106].

For the beta-beam facility studied in [86] the lattice and aperture have been chosen so that lost ions can be dumped in a controlled way or intercepted by absorbers [107]. The decay products can successfully be kept in the machine along the straight sections and extracted to a beam dump before the arc with the help of a dipole and classical septa magnets (see Fig. 2.13). The arcs can be specially designed for used ions. For the beta-beam facility studied in [86] the arc lattice have been derived for $^{18}$Ne and $^6$He and large aperture dipoles (160 mm diameter) are used with such a length that all lost ions can be intercepted with absorbers after the dipole in which the losses occur or after the consecutive dipole [107, 108]. The choice of absorber length and material is important as the absorbers – in the worst case – could simply serve as a point for the ion to break up in a hadron cascade which could heat the following (superconducting) dipole badly. To find the right design, a combination of particle tracking and

Horizontal Plane

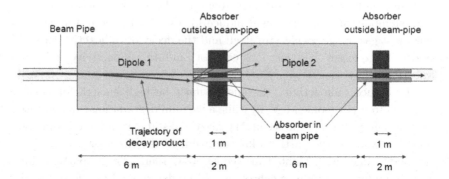

Fig. 2.12   The beam loss from radioactive decay in the arcs cannot be avoided and the dipoles must be adapted in length so that the daughter products are lost in absorbers between the dipoles and not in the dipoles themselves. Illustration courtesy of Elena Wildner at CERN.

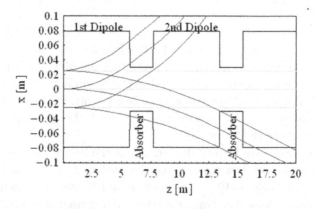

Fig. 2.13   For the beta-beam facility studied in [86] the length of the dipoles in the arcs have been chosen [107] so that the lost lithium ions (from decay of $^6$He) are lost between the magnets and not in the magnets where they could damage or overheat the superconducting coils. Illustrations courtesy of Jacques Payet and Antoine Chance at CEA.

matter interaction codes had to be used [110]. This work also demonstrates that the overall losses in a representative part of the arc are kept under the quench limit for classical superconducting dipoles but that at certain points in the superconducting coils the losses are likely to cause quenches. A possible way forward is to develop open mid-plane superconducting magnets in which the dominant losses in the horizontal plane of the beam are absorbed in the cold mass of the magnets, and not in the sensitive superconducting strand [111]. Another solution is to use a thick liner inside the dipole which would distribute the decay products over a larger volume of the magnet mass and coil. The daughter ions from the decay in the straight sections can be transported together with the ion bunches up until the first dipole in the arc where they can be separated and dumped in a controlled manner [107] (see Fig. 2.14).

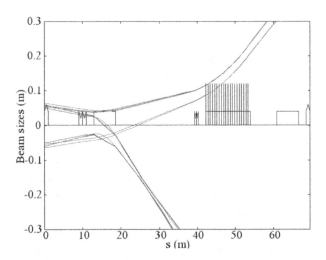

Fig. 2.14   For the beta-beam facility studied in [86] the daughter ions from the decay in the straight sections are transported along the straight section together with the original ion bunches and steered to beam dumps at the beginning of the arc [107] to reduce the activation of elements and rock in the decay ring. Illustrations courtesy of Jacques Payet and Antoine Chance at CEA.

## 2.6   Stacking and Storage

### 2.6.1   *Why do we need stacking?*

At low neutrino energies the background signal in the detector from atmospheric neutrinos will be a major issue. To enable an efficient suppression of this background it is necessary to operate the beta-beam facility with a low duty factor where the duty factor is defined as

$$\text{Duty factor} = \frac{Nt_b}{T_{rev}} \tag{2.2}$$

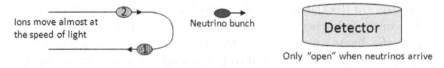

Fig. 2.15   The neutrino beam from the beta-beam facility will have a bunched structure
which mirrors the bunch structure of the ion-beam. In the figure, the neutrino bunch
originates from the ion bunch marked 1, the neutrinos from the decay of ions in the
bunch marked 2 have not left the ring yet but are travelling along the straight section
with the same velocity as the ions.

where $N$ is the number of filled bunches in the decay ring, $t_b$ is the length
of the individual bunch in seconds and $T_{rev}$ the revolution time. The fact
that the neutrino beam will mirror the bunch structure of the ring is not
obvious at first glance but follows from the fact that the ion bunches travel
at a speed very close to that of light. The neutrinos emerging from the
beta-decay process in one straight section will simply accompany the ion
bunches up until the first bending magnet where the neutrino bunches
will continue straight forward while the ions follow the closed orbit of the
ring (see Fig. 2.15). The limits for an acceptable duty factor are set by
several parameters and will be discussed in detail in Section 3.3.1.3. For
our discussion in this chapter it is sufficient to know that the duty factor
for a beta-beam facility operating at Lorentz $\gamma = 100$ using isotopes with
an average neutrino energy spectrum of a few MeV at rest would require
a duty factor not larger than a few $10^{-3}$. The lifetime of an isotope with
1 second half-life at rest will be 100 seconds at $\gamma = 100$. The beta-beam
complex discussed in Section 2.2 will accelerate one batch of ions every 5–10
seconds, so to keep the duty factor low and the intensity high it is desirable
to put fresh ions into the same buckets in which the already circulating ions
reside. Dumping the circulating beam which has a radioactive half-life of
some 100 seconds after only 5–10 seconds for a refill would be a waste of
ions, and to increase the time between refills would also result in a lower
rate of neutrinos (see Fig. 2.16). To do this, it is necessary to perform
some form of stacking, and if possible, also some form of beam cooling.

### 2.6.2   Beam-cooling

Cooling in the context of particle beams usually refers to the compression
of phase space, either transverse phase space which is spanned by beam
size and divergence, or longitudinal phase space which is spanned by bunch

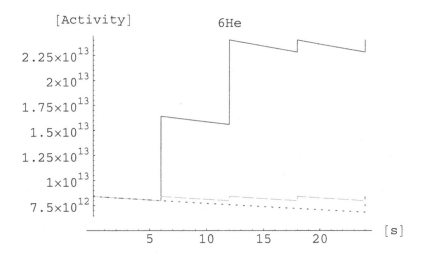

Fig. 2.16  The radioactivity in the decay ring for the beta-beam facility studied in [86] using: i) a stacking scheme with up to three injection pulses stacked before losses stabilizes the achievable activity (full line), ii) a re-filling scheme with the same frequency as for stacking but in which the circulating bunch is killed before each injection (dashed line) and iii) a re-filling scheme with a lower frequency to permit more of injected ions to decay before the bunch is killed (dotted line).

length and energy spread. Phase space is referred to as the *emittance* ($\epsilon$), for the transverse plane it is measured in units of $\pi m \times mrad$ and longitudinal case in units of $eV \times s$. The total phase space available will for the transverse planes be set by the available machine aperture and the machine lattice; for the longitudinal case the available RF voltage and frequency of the RF system. The important point is that there is a **limited** amount of phase space available in all dimensions and that given a certain value for the incoming beam emittance it will only be possible to fit a certain number of beam batches into this phase space. The only way to increase the space available for a new beam is by compressing the particles which already are in the decay ring into a smaller part of phase space. We refer to this process in accelerator physics as beam cooling.

There are two different techniques for beam cooling in synchrotrons, electron cooling and stochastic cooling. The techniques differ in the way they are applied but the conclusion for beta beams is that for ions with a $\gamma$ above 100, the cooling times for any realistic set of cooling system parameters measure in tens of minutes which is too long to have any practical use in stacking.

### 2.6.3   *Stacking*

The injection process used between the linac and the first synchrotron in
the beta beam scenario in [86] is a form of transverse stacking. For the
decay ring it would be impractical to use this method as the time between
injections into the decay ring is long and the beam is bunched. A more
useful approach is to stack in longitudinal phase space [109].

To perform longitudinal stacking two RF systems are required with one
system having half the frequency of the first. In addition, fast switching
magnets are needed for the injection of the new ion bunches. In the first
step the ion bunches are injected at a slightly higher (or lower) energy than
the circulating ions, which will force the beam to follow a different orbit of
a larger radius in the machine. In this orbit the injection system with the
fast switching magnets will not interfere with the circulating ions which is
necessary as the magnets would be too slow to act between machine bunches
and could accidentally disturb, or even eject, the circulating beam. The
injected bunches are in the next step grabbed by the RF field of a cavity
of half the nominal frequency but with a high voltage and accelerated (or
in longitudinal phase space rotated) to the nominal beam energy, thereby
to the same orbit as the circulating ions. In the last step the smaller
incoming bunch is merged with the larger circulating bunch using the two
RF systems and adjusting amplitude and phase in such a way that only the
small incoming bunch is asymmetrically merged with the center part of the
large circulating bunch.

The net effect is that phase space density in the center of the circulating
bunch is increased and that the outer parts of the bunch (the edges of the
bunch and the particles with largest energy diversion) are pushed outside
the potential well formed by the RF system and lost. The process has been
simulated using the ESME code [112] and in Fig. 2.17 the four major stages
of the process are shown.

The feasibility of the asymmetric bunch merging process was also tested
in the CERN PS with very good results. In the PS tests empty phase space
was merged with very high efficiency with a high intensity proton bunch
in the PS (see Fig. 2.18). For the beta-beam facility used as a reference
in this chapter the merging can be done up to 15 times for an $A/Q$ of 3
and up to 20 times for an $A/Q$ of 2. The RF system makes use of an
80 MHz and a 40 MHz RF system. The increase in number of possible
merges before filling phase space with the inverse of the $A/Q$ is due to the
increased longitudinal focusing at lower $A/Q$s. In reality the efficiency of

a        b        c        d

Fig. 2.17 The stacking scheme proposed for a low-duty-cycle beta-beam facility has been simulated with ESME. The simulation is done after the stacking reaches equilibrium and starts (a) with the injection of a "fresh" parallel in phase (or time) to the circulating particles but at a higher energy and consequently in a different orbit. The next stage (b) is the rotation with RF cavities of the bunches so that they both end up at the same energy. This is followed by the asymmetric merging (c) of the center of the large circulating bunch and all of the injected bunch. In the last picture (d) the resulting circulating bunch can be seen to have an onion-like structure due to the sequential merging of the center part which pushes ions which have been circulating longer towards the edge and eventually out of the bucket. Illustration courtesy of Steven Hancock at CERN.

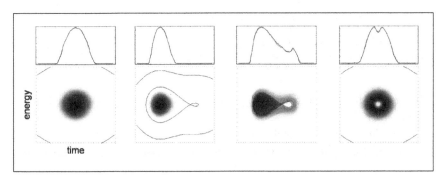

Fig. 2.18 The stacking scheme was tested in the CERN PS [109] where a high intensity proton beam was merged with a small part of empty phase space with high efficiency. In the top part of the figure is the line current for the bunches during the merging process and in the bottom part the tomographic [113] reconstruction of the bunches are shown. Illustration courtesy of Steven Hancock and Michael Benedikt at CERN.

the merging will be further limited by the phase stability of the RF system and assuming a realistic RF parameters and stability an overall stacking efficiency of 80% can be reached for both $^6$He and $^{18}$Ne [114]. In Fig. 2.19 the efficiency of each stacking cycle is plotted as a function of number of merges. The ions which are pushed out of the potential well formed by the RF cavity will eventually be lost. The total beam power injected is about 1 MJ for the reference facility and up to 50% of that beam power is lost from

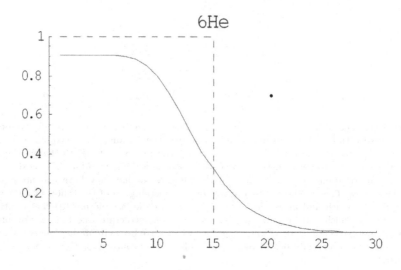

Fig. 2.19   The stacking efficiency for each merge will decrease as the RF bucket in the decay ring gets filled. The efficiency is plotted as a function of number of merges (maximum 15 for $^6$He), the dashed rectangle is the ideal case in which 100% efficiency is achieved for each merge until the RF bucket is full after 15 merges for $^6$He. Illustration courtesy of Steven Hancock at CERN.

the bucket during stacking and has to be "scraped" away with longitudinal collimators.

The decay ring [107] will have to be specially designed for this stacking scheme with a large energy spread in the transverse plane – or large dispersion as accelerator physicists call it – for the injection region and another large dispersion region for beam collimation. The "natural" dispersion in arcs caused by the bending dipoles can be used for the injection while the collimation is better done in one of the straight sections. The reason for this is that the requirement to keep the decay ring compact and the arcs short compared to the straight sections will require the use of superconducting magnets which are sensitive to beam loss. The collimation is done with scrapers – or collimators – which will catch the ions pushed out of the bucket by the merging process and consequently have the highest energy deviation from the central beam energy. The dispersion at the collimators will ensure that only ions with high energy deviation are intercepted by the transversely acting collimators.

A further consideration for the decay ring design is the average (optical) beam divergence in the straight sections. It can be estimated from the beam emittance:

$$D_{average} = \frac{\epsilon_{h,v}}{\beta_{average}} .$$

### 2.6.4 *Annual rate of neutrinos*

The purpose of the beta-beam facility is to produce a well collimated beam of single flavor electron (anti)neutrino beams. For the ideal beta-beam facility in which all ions accelerated can be stored in the decay ring until they produce a neutrino and are lost, the annual rate is simply the number of ions injected per time unit multiplied by the length of the running period per year in seconds. In Particle Physics the typical running period in one year is called a *Snowmass year* – named after the recurrent particle physics meeting in Snowmass in the USA [115] – which is $10^7$ seconds long. This is roughly a third of a real year and it simply reflects the typical length of a particle physics run at a large accelerator laboratory.

When the decay ring only can stack a certain number of bunches the upper intensity limit in the machine can be calculated from the truncated series of repeated injections in the decay ring. In Fig. 2.20 the actual fraction of stored ions compared to the ideal case in which all injected ions are stored in the ring is plotted as a function of the number of "merges" for both $^6$He and $^{18}$Ne. Note that the realistic case of only 15 merges for $^6$He and 20 merges for $^{18}$Ne only corresponds to an efficiency of the stacking scheme of 54% and 26% respectively.

The maximum number of surviving ions from $n$ repeated injections of $N$ radioactive ions into the decay ring every $T$ seconds can be written as

$$N_{tot} = N + Ne^{-\lambda T} + Ne^{-2\lambda T} + ... + Ne^{-(n-1)\lambda T} \qquad (2.3)$$

where $\lambda$ is the disintegration or decay constant of the isotope in question. Using the geometric series Eq. (2.3) can be written as:

$$N_{tot} = N\frac{1 - e^{-\lambda T n}}{1 - e^{-\lambda T}} . \qquad (2.4)$$

The final result converges for an infinite number of injections at an expression which simply states that the number of decays at equilibrium will equal the number of ions injected. For the realistic stacking scenario the maximum number of useful decays producing neutrinos is truncated. Note that this is the **maximum** number of useful decays as it assumes perfect stacking from the center of the bunch.

The decay constant for the radioactive decay producing neutrinos in the right energy interval is not necessarily the only decay constant determining

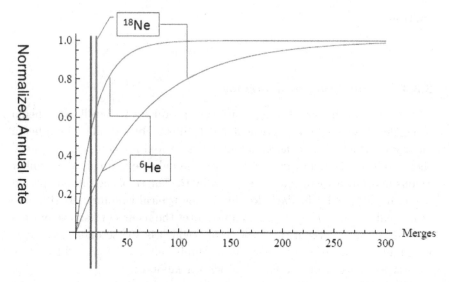

Fig. 2.20   In the diagram the fraction of $^6$He ions (upper line) and $^{18}$Ne ions (lower line) stored in the decay ring using the stacking scheme discussed in Subsection 2.6.3 are plotted as a function of the number of 100% efficient merges. The ideal case to which the curves eventually converge is the ideal case in which all ions are accumulated in the ring until they decay. The vertical lines for each ion type mark the stacking limits for the beta-beam facility studied in [86], 15 merges for $^6$He and 20 merges for $^{18}$Ne.

the decay-rate of the ions in the decay ring. There might be other loss processes such as vacuum collisions – resulting in a change of mass-to-charge ratio of the ion – which can be expressed as a decay constant and which dominate the total decay rate. For monochromatic neutrino beams from electron capture decay there is often a radioactive branch of competing $\beta+$ decay with a different decay constant. At stable operation, with the maximum total number of ions in the ring being constant, the number of "useful" decays from the channel p1 can be calculated as the integral of the activity $\lambda_{p1}N_{tot}e^{-\lambda_{all}T}$ for one stacking period, $T$. Just to avoid any misunderstanding, the subscript *p1* refers to the partial decay constant for the decay channel producing the "useful" neutrinos and *all* to the sum of all decay constants ($\lambda_{all} = \lambda_{p1} + \lambda_{p2} + \ldots$). For radioactive decay the decay constant will be time dilated and relates to the decay constant at rest as $\lambda = \lambda_{rest}/\gamma$.

For later use in the discussion on storage of partly stripped nucleus or nucleus with more than one decay channel it is helpful to write out the resulting integral and its general solution.

$$A = \frac{1}{T} \int_0^T \lambda_{p1} N_{tot} e^{-\lambda_{all} t} dt = \frac{\lambda_{p1} N_{tot}}{T} \left[ \frac{e^{-\lambda_{all} t}}{-\lambda_{all}} \right]_0^T = \frac{\lambda_{p1} N_{tot}}{\lambda_{all} T} (1 - e^{-\lambda_{all} T}).$$

$$(2.5)$$

For any form of beta decay in which one decay process (p1) dominates $(\lambda = \lambda_{all} \simeq \lambda_p)$ and with $N_{tot}$ from Eq. (2.4)

$$A = \frac{N}{T} \left( 1 - e^{-\lambda n T} \right). \qquad (2.6)$$

To get the annual rate at the end of one of the straight sections for a (Snowmass) year of $10^7$ seconds we have to multiply 2.5 by the relative length of the straight section ($f$) compared to the circumference of the ring which finally gives the annual rate of neutrinos:

$$A = 10^7 f \frac{N}{T} \left( 1 - e^{-\lambda n T} \right). \qquad (2.7)$$

The annual rate of the beta-beam facility studied in [86] is assumed to reach $1.1 \times 10^{18}$ electron neutrinos per year from the decay of $^{18}$Ne and and $2.9 \times 10^{18}$ electron antineutrinos per year from the decay of $^6$He. To reach thess ambitious goals a production rate of $2 \times 10^{13}$ ions per second of each species is required. The overall efficiency from the ion source to the number of ions injected into the decay ring is assumed to reach 12% for $^{18}$Ne and 25% for $^6$He. Using the previously quoted efficiencies for the stacking scheme the total efficiency from the ion source to ions decaying in one straight section (which is 36% of the total circumference) is 1% and 5% respectively.

### 2.6.5 *Other limitations*

The acceleration of particles in a synchrotron requires an RF system which can deliver a certain power to the beam for acceleration. The efficiency in the coupling between a given RF system in a given ring is frequency dependent and is *e.g.* for the CERN PS 20%. The PS was built for fixed target physics with protons which are easy to produce and ionize. Consequently, the total number of charges which can be handled is an impressive $4 \times 10^{13}$, sufficient for both ion species in the beta beam studied in [86].

The aperture of any accelerator will evidently limit the size of the injected beam. Together with the lattice of the machine this translates into a certain acceptance which is calculated in the same units as beam emittance. An obvious drawback of using an existing machine is that the acceptance is difficult to change and if the machine has been designed for low emittance beams it might simply be useless for a high intensity ion beam. Just as for the RF power limitations the study in [86] found that the similarities between protons and ions and the fact that the CERN synchrotrons were designed for a high intensity fixed target physics program makes it possible to reach a reasonable performance using existing CERN machines for the beta beam.

## 2.7   Possible Future Development

The beta-beam facility first studied at CERN in 2002 [86] was built with a requirement to re-use a maximum of the existing accelerator infrastructure and to – as far as possible – only depend on known technologies for e.g. production and bunching. It is possible to imagine a rather different facility if these constraints were ignored. It is apparent from physics reach studies that if the annual rate of the neutrinos could be increased to some $10^{19}$ electron (anti)neutrinos per year, the beta beam concept would have a much larger scope. In this section some ideas which go beyond the original CERN baseline are discussed.

### 2.7.1   *Accumulation at low energy*

The magnetic field in a synchrotron has to be increased during acceleration, making it impossible to accelerate a continuous beam. New particles can only be injected once the magnets are back to the field corresponding to the injection energy. The time between two injections can be as long as several seconds for a high energy synchrotron such as the PS and SPS at CERN or as short as 20 milliseconds for a rapid cycling synchrotron such as the ISIS at Rutherford labs. The combination of synchrotrons proposed for the earlier discussed CERN beta-beam facility will induce a total dead-time of up to almost ten seconds for the production side. The simplest way to make use of this lost production time is to accumulate the produced ions before further acceleration (see Fig. 2.21). The accumulation can in principle be done at rest in some form of electromagnetic trap e.g. an ECR source with a long retention time. However, the more common solution is to use a low energy storage ring with a beam cooler to accumulate and cool the ions. Such a scheme is used for the CERN Large Hadron Colliders (LHC) ion physics programme. The acceleration time is used to accumulate and cool

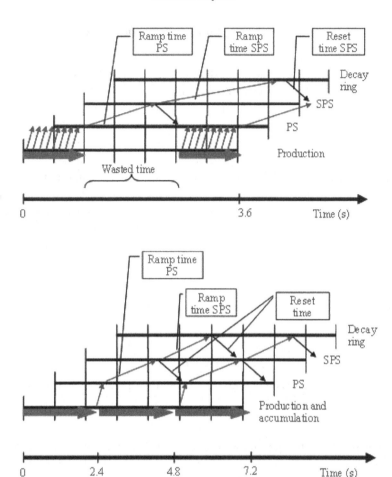

Fig. 2.21  Accumulation at low energy is a gain for the annual rate of neutrinos from the facility as radioactive ions can be produced and stored while acceleration of an earlier production batch is in process. The figure shows the machine cycle without accumulation (top) and with accumulation (bottom).

intense and small bunches of lead ions to achieve a reasonable luminosity (collision event rate) in the LHC detectors. A study has been done for the beta beam in a similar set-up [116]. The main difference to the LHC type accumulation ring is that the ions are radioactive and will decay. The result is that it only makes sense to accumulate for up to some three half-lives in total. However, the gain in intensity for the neutrino beam can be up to a factor of five, significant for isotopes which are difficult to produce.

### 2.7.2    *Two isotopes in the ring at the same time*

A beta-beam facility can produce both the electron neutrino and the anti-electron neutrino from beta$^-$ and beta$^+$ decaying ions. In principle, the two types of ions can be kept at the same time in the decay ring which, in theory, would make it possible to make a measurement in a shorter time. The ions would be kept in different bunches sufficiently well separated in time to permit the experiment to treat the signal from either ion type independently. To keep ions with different charge ($q$) and mass ($m$) on orbit in a ring with a given bending radius ($\rho$) and a fixed magnetic field ($B$) in the dipoles, the momentum ($P$) of the two ion types must be different. For our purpose where the energy is high (Lorentz $\beta \approx 1$ for both ion types) and the key parameter for the performance of the facility is the kinematic boost of the ions, it makes more sense to discuss the relationship between the Lorentz $\gamma$ of the two ions rather than the relationship between the momenta. Given that

$$\frac{P}{q} = \frac{\gamma \beta m}{q} = B\rho$$

and

$$\beta_A \approx \beta_B \approx 1$$

the relationship between the Lorentz $\gamma$ factors of the ions becomes

$$\gamma_A = \frac{m_B/q_B}{m_A/q_A} \gamma_B .$$

For a symmetric storage ring with an average bending radius ($\rho$) in the arcs and with two equally long straight sections ($L$) the revolution time can be written as

$$T_{revolution} = \frac{2\pi\rho + 2L}{\beta c} .$$

The revolution time in the ring for the two ion types must be identical to enable longitudinal control of the beam with a common RF system for storage and accumulation. For two types of ions (A and B) this requirement will induce a difference in average radius ($\rho$) between the ion types to compensate for the difference in $\beta$. For

$$\rho_A = \rho_B + \Delta\rho$$

the difference becomes

$$\Delta\rho = \frac{\beta_A}{\beta_B} \left( \frac{L_B}{\pi} + \rho_B \right) - \left( \frac{L_A}{\pi} + \rho_B \right) .$$

Note that the index (A and B) for the straight sections refers to the ion types rather than the two different straight sections. The fact that the ions have slightly different orbits can be used to separate them in the straight sections so that one ion type is made to follow a longer beam path (e.g. by adding a small loop to the straight section for one ion type) which would reduce the average radius difference. However, there are several problems with keeping two ion types in the machine. One is that the radius difference for the typical gamma of beta-beam facilities will be sufficiently important to require a large horizontal aperture of the dipole magnets which can be very costly. Another is that the production, acceleration and stacking also must be doubled if there should be any real gain for the experiment. The main bottleneck in present proposals is the acceleration cycle which would have to be repeated twice and in sequence for two different ion types. As this would take twice the time, the loss in accumulation rate for each ion type would lead to a lower annual rate for each neutrino type and no overall gain for the experiment. Note that even if this problem could be partly overcome with an accumulation stage at low energy and some doubling of the accelerator stage, the accumulation and merging in the decay ring into a few bunches to keep the duty factor low would be extremely complicated, maybe even impossible due to the shared RF system.

### 2.7.3 *Higher gamma*

Several physics reach studies have been done for a beta-beam facility with higher energy than the one proposed in [86]. The consequences for the machine are important, not least the fact that the existing accelerators at CERN which formed an essential part of the accelerator complex in [86] only can accelerate $^{18}$Ne to a gamma of 250 and $^{6}$He to a gamma of 150. Still, other laboratories have synchrotrons which can accelerate to higher gammas e.g. a conceptual study was done at Fermi lab for a gamma of 300 using the Tevatron as the pre-accelerator (see Fig. 2.22).

There are consequences for the focusing of the neutrino beam with an increase in gamma as the focusing to to first order is inversely proportional to gamma. The opening angle, $\Theta$, of the neutrino beam as given by the kinematic focusing is: $\Theta \approx 1/\gamma$. For the decay ring design the most important differences are a) that the life-time will be longer due to increased time dilatation which will influence the stacking efficiency and the annual rate at the end of the straight section (see Figure 2.23) and b) that the decay ring will have to be larger or the dipole magnets more powerful to

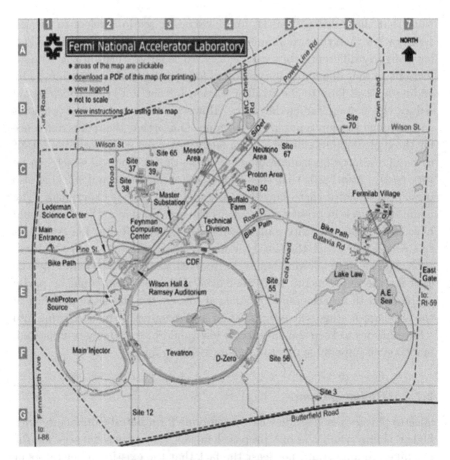

Fig. 2.22   A beta beam decay ring with a magnetic rigidity of 1500 Tm and a gamma of 300 would fit on the present Fermi laboratory site. Illustration courtesy of Andreas Jansson from FNAL.

cope with the increased magnetic rigidity of the radioactive ions. Assuming a perfect arc completely filled with dipoles the length of the decay ring can be calculated for different gamma (see Table 2.4).

### 2.7.4   *Barrier buckets in the decay ring*

At neutrino energies corresponding to atmospheric neutrinos it is very important to keep the duty factor low to permit suppression in the experiment of atmospheric background. The result is that only a fraction of the capacity of the decay ring can be efficiently used as only a limited number

Table 2.4 Some possible decay ring options for a different Lorentz gamma of $^6$He. The decay ring arcs are in all cases considered to be completely filled with dipoles.

| Gamma | Rigidity [Tm] | Ring length[a] | Dipole field[b] |
|-------|---------------|----------------|-----------------|
| 100 | 935 | 4197 | 3.1 |
| 150 | 1403 | 6296 | 4.7 |
| 200 | 1870 | 8395 | 6.2 |
| 350 | 3273 | 14691 | 10.9 |
| 500 | 4676 | 20987 | 15.6 |

[a]Assuming a fixed field of 5 T and a single straight section of 36% of the circumference.
[b]Assuming an arc radius of 300 m and a decay ring length of 6885 m.

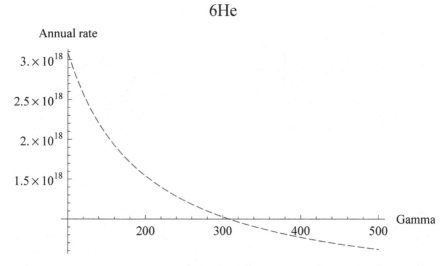

Fig. 2.23 The annual rate as a function of the $\gamma$ in the decay ring for $10^{14}$ $^6$He ions stored in the ring. One of the straight sections of the ring represents 36% of the total circumference.

of injected pulses can be accumulated in a single bunch (see [109]). To make full use of the storage capacity of the decay ring the beam could be kept unbunched in the decay ring. The problem with an unbunched beam which fills the full circumference of the decay ring is that it is impossible to inject without disturbing the beam in the ring. A possible solution is to use RF cavities as barriers for the unbunched beam so that an "injection hole" is created for the new beam from the injectors. This kind of longitudinal beam manipulation in a storage ring is referred to as "barrier buckets" and has been tested with high intensity proton beam at the AGS in Brookhaven [117] (see Fig. 2.24). The consequence of this is of course that the neutrino beam will have no real duty cycle. The injection hole in the beam will create some empty time slot in the neutrino beam which maybe could serve as a reference for background estimates in the detector.

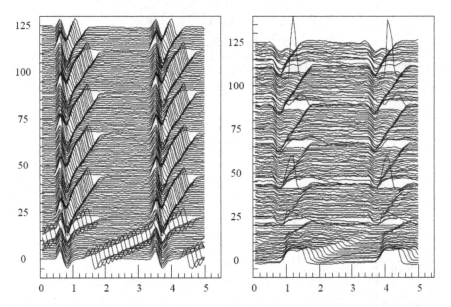

Fig. 2.24 If the physics requirements on the duty factor can be relaxed, a "barrier bucket" stacking scheme can be used such as the scheme tested at the Brookhaven AGS for accumulation of protons [117]. In the left part of the figure the voltage of the two cavities acting as "barriers" are shown versa time in the ring for different time slices of the accumulation process. In the right part of the figure the corresponding proton beam intensity is shown versa time. Note how the two RF barriers move apart to create an injection hole in the otherwise continuous beam of the ring.

### 2.7.5 *Acceleration of partly stripped ions*

Proton-rich nuclei can also decay via electron capture and the neutrino emitted will in this decay mode be mono-energetic as there is no positron emitted simultaneously. The electron capture process is often the only possible decay mode close to stability where there is insufficient energy available in the decay to form the required electron-positron pair for $\beta^+$ decay.

The lifetime of most isotopes decaying with electron capture is generally long which makes it difficult to use this decay mode for the production of a mono-energetic electron neutrino beam [118, 119]. The exceptions are some exotic rare-earth isotopes [120] in which the decay to the ground state in the daughter nuclei is highly hindered so that the electron capture process to a higher lying excited state can compete. The equivalent process on the neutron-rich side is bound beta-decay in which the emitted electron is captured in an atomic orbit and the anti-electron neutrino is emitted with a definite energy. The branching rate for this process is generally very small but there has been a proposal to use it in combination with electron capture decay for a CP-violation measurement [121].

A definite requirement for electron capture decay is that the nucleus only is partly stripped so that there is an electron available for capture. The acceleration of partly stripped nuclei is fairly straightforward [122]. The main difference compared to accelerating fully stripped radioactive ions is that the loss (or gain) of electrons will change the mass-to-charge ratio of the isotope which will cause additional losses. At high energy the likelihood to pick up an electron is vanishingly small. However, the likelihood of losing an electron will, expressed in an equivalent half-life, be in the order of minutes in a ring with a modern ultra high vacuum system. The annual rate for electron capture decaying nuclei is modified by the "vacuum halflife" which will compete with the radioactive decay in the straight section but without producing any neutrinos. If the two processes are expressed as decay constants the total decay constant can be written as the sum of the two, $\lambda_{tot} = \lambda_{vac} + \lambda_{ec}$. Inserting this in the integral in Eq. (2.5) the modified annual rate (for a Snowmass year of $10^7$ seconds) can be written as

$$Rate = \frac{Nf}{T} \times \frac{\lambda_{ec}/\gamma_{top}}{\lambda_{ec}/\gamma_{top} + \lambda_{vac}} \times \left(1 - e^{-mT(\lambda_{ec}/\gamma_{top} + \lambda_{vac})}\right) \times 10^7$$

where $N$ is the number of incoming ions into the decay ring per injection cycle, $f$ the fraction of the decay ring being a straight section pointing towards the detector and $T$ the time between injection cycles.

The half-life of the partly charged electron capture decaying nuclei with only n $s$ electrons left will scale as $1/n^2$ [123, 124] e.g. a Dy atom with only one $1s$ electron left would still yield 40% more than 40% a neutral Dy atom.

The isotope $^{152}Tm$ with a half-life of 8 seconds is one of the shortest living nucleus with an important part of the decay going via electroncapture. This is still a half-life five times longer than $^6$He which will have a negative influence on the annual rate for the same amount of isotopes stored in the decay ring. This rather heavy nucleus with 69 protons ($Z = 69$) would typically have a charge state of above 50 at higher energies. The combination of the high charge state, the longer half-life and the electron stripping losses will require a large number of ions to be accelerated and stored in the decay ring to keep the annual rate high. For this specific case the tune shift in the CERN accelerators PS and SPS would peak well above 0.25 to keep the annual rate at $10^{18}$ electron neutrinos at the end of one straight section for a year of $10^7$ seconds.

# Chapter 3

# CERN-Fréjus Beta Beam Physics Potential

## 3.1 Introduction

A beta beam is produced from the decay of a high energy radioactive ion beam, resulting in a pure $\nu_e$ or $\bar{\nu}_e$ beam. The flavor transitions that can, in principle, be studied in this facility are:

$$\nu_e \to \nu_\mu \quad \nu_e \to \nu_e \quad \nu_e \to \nu_\tau$$
$$\bar{\nu}_e \to \bar{\nu}_\mu \quad \bar{\nu}_e \to \bar{\nu}_e \quad \bar{\nu}_e \to \bar{\nu}_\tau.$$

In the laboratory frame, the neutrino flux, $\Phi^{\text{lab}}$, is given by [125]:

$$\left. \frac{d\Phi^{\text{lab}}}{dSdy} \right|_{\theta \simeq 0} \simeq \frac{N_\beta}{\pi L^2} \frac{\gamma^2}{g(y_e)} y^2 (1-y) \sqrt{(1-y)^2 - y_e^2} \tag{3.1}$$

where $N_\beta$ is the number of ion decays per unit time, $Q_\beta$ is the endpoint kinetic energy of the beta particle, $\gamma$ is the relativistic Lorentz boost factor, $m_e$ is the mass of the electron, $dS$ is the element of solid angle, $L$ is the distance between the decay ring and the detector, $0 \leq y = \frac{E_\nu}{2\gamma Q_\beta} \leq 1 - y_e$, and $y_e = m_e/Q_\beta$; and

$$g(y_e) \equiv \frac{1}{60} \left\{ \sqrt{1 - y_e^2}(2 - 9y_e^2 - 8y_e^4) + 15y_e^4 \log \left[ \frac{y_e}{1 - \sqrt{1 - y_e^2}} \right] \right\}. \tag{3.2}$$

The intensity and the energy shape of the neutrino beam are determined by just four quantities: $N_\beta, Q_\beta, \gamma, L$. Once these parameters are fixed, the neutrino flux can be calculated precisely since the kinematics of $\beta$ decay is very well-known [127].

There are some approximative scaling laws at the varying of the parameters (assuming $N_\beta$ constant): the maximum $\gamma$ to which a given accelerator can accelerate a ion is proportional to $Z/A$. For instance, if SPS can accelerate $^6$He $(Z/A = 2/6)$ up to $\gamma = 150$, $^{18}$Ne $(Z/A = 10/18)$ can be accelerated up to $\gamma = 250$.

The neutrino flux $\Phi$ at a far detector placed at a distance $L$ is:

$$\Phi \propto \frac{\gamma^2}{L^2}$$

because the emission angle of the neutrino from the parent ion, in the laboratory frame, is proportional to $\gamma^{-1}$.

Since the optimal distance $L$ is defined by the oscillation $\Delta m^2$: $L \propto E_\nu / \Delta m^2$ and $E_\nu \propto \gamma Q_\beta$ the flux becomes

$$\Phi \propto \frac{(\Delta m^2)^2}{Q_\beta^2}.$$

Considering that the neutrino interaction rate $I$ at the far detector is $I = \sigma \Phi$ and that the neutrino cross section $\sigma$ goes as $\sigma \propto E_\nu$ (this scaling law becomes inaccurate for $E_\nu < 5$ GeV) we can derive the important merit factor $\mathcal{M}$

$$\mathcal{M} \propto \frac{\gamma}{Q_\beta}. \tag{3.3}$$

It follows that performances of a beta beam scale as the Lorentz boost factor $\gamma$ and are inversely proportional to the endpoint energy $Q_\beta$.

It should be noted that the end point energy of a muon decay being 68 MeV while a suitable beta-decay isotope as $^6$He has an end-point energy of about 3.7 MeV, the merit factor of a beta beam is about 20 times better than the merit factor of a neutrino factory.

Besides these scaling laws there are other very important considerations to be taken into account for the choice of ions and of $\gamma$, as discussed in Section 2.3.1, such as the production rate and the lifetime of the ions.

Of course the above discussion does not take into account any detector considerations, that could heavily affect overall performances because of detection thresholds, signal efficiencies and detector backgrounds.

## 3.2   The CERN-Fréjus Configuration

The CERN beta beam can accelerate $^6$He ions up to $\gamma = 150$ and $^{18}$Ne ions up to $\gamma = 250$. Given the characteristics of the $^6$He decay, this translates to mean neutrino energies of up to $\sim 600$ MeV, equivalent to a maximum baseline of 300 km.

The only realistic candidate site for the excavation of a megaton class detector fitting this request is the Fréjus site, at a distance of 130 km.

To fit this distance the optimal $\gamma$ for $^6$He is $\gamma \simeq 100$. Higher $\gamma$ values would increase interaction rates in the detector, but not the oscillated event interaction rates by very much, since the baseline would no longer fit the oscillation pattern. Furthermore background rates would rise, as discussed in Section 3.3.1.

Smaller $\gamma$ values would have the advantage of suppressing background rates in the detector, $\gamma_{^6\text{He}} = 66$ had been indeed the initial choice for the CERN-Fréjus configuration [128] for this reason. Under this condition however the neutrino flux is smaller and a bigger fraction of $\nu_\mu$ events created by oscillations produces a muon below the Čerenkov light production threshold ($p_\mu > 120$ MeV/c).

The CERN-Fréjus configuration (CFBB) is not designed to be the absolute optimal configuration for a beta beam experiment. It is intended to be a realistic setup where both the beam and the detector sites are chosen among realistic conditions.

## 3.3 Data Analysis

The most sensitive process in a beta beam experiment are $\nu_e \to \nu_\mu$ transitions as will be discussed in Section 3.4.

They introduce an experimental problem never faced so far, the detection of a small content of $\nu_\mu$ events in a pure $\nu_e$ beam. This process can be complemented by $\nu_e \to \nu_e$ transitions, where a small deficit in $\nu_e$ spectrum is looked for.

The combination of the two processes demands a detector capable of measuring with precision and high purity both electrons and muons. Furthermore, as we will see, to achieve good sensitivities for leptonic CP violation, the detector should be massive, in principle several units of 100 kt.

The water Čerenkov technology, following the extremely successful experience of Super-Kamiokande, is the default choice for such a detector. Liquid argon or liquid scintillator detectors are in principle also good candidates, but their cost per unit mass is much higher than water. These detectors can be competitive only for energy regimes where the water Čerenkov technology becomes less efficient because multi-ring events become the dominant process (see the discussion in Section 1.3.4.6), this energy is around 1.5-2 GeV.

The main problematics of this kind of experiment and the different experimental approaches that have been proposed to attack the problem will be discussed in the next sections.

### 3.3.1 Backgrounds

While a beta beam provides an absolutely clean beam of $\nu_e(\overline{\nu}_e)$, backgrounds can be produced by imperfect performances of the neutrino detector. The general problems will be discussed in the following, together with a quantitative analysis of the specific case of the CERN-Fréjus setup. The main sources of backgrounds are:

- $\nu_e$ interactions in the detector where the outgoing electron is identified as a muon;
- neutral-current (NC) interactions where a charged pion is produced and then identified as a muon;
- atmospheric neutrino interactions in the detector, producing genuine $\nu_\mu$ charged-current events.

#### 3.3.1.1 Backgrounds from $\nu_e$ interactions

The experimental sensitivity requires that the electron-muon misidentification rate in the detector must be kept below $10^{-4}$. It is extremely demanding to keep this rate so low.

A water Čerenkov detector is particularly efficient in this aspect, basing its rejection on two powerful handles. First muon and electron events have very different topologies in the detector, the former producing a rather sharp ring, the latter a rather fuzzy ring. Furthermore a muon can be positively identified by detecting its decay products: a Michel electron of energy up to $m_\mu/2$ with a characteristic time delay given by the muon lifetime.

The Super-Kamiokande collaboration already developed these techniques, demonstrating that the electron-muon identification can be kept below $10^{-4}$ for particle momentum below 1 GeV.

A Michel electron is produced only when a muon arrives to decay, a process in competition with the muon absorption in water. This is the reason why neutrino interactions are less efficiently detected than antineutrino interactions (positive muons are repulsed and not attracted by nuclei). This is also the reason why in heavier Z targets, for instance argon, this feature cannot be used, as the probability for a negative muon to be absorbed before its decay is too high.

### 3.3.1.2 *Backgrounds from neutral-current interactions*

The charged pions produced in the process

$$\nu N \to \Delta \to N\pi$$

where $N$ is a generic nucleon and the charges are not specified, can be mis-identified as muons, generating backgrounds. At the energies typical of a beta beam, the momentum of these pions is such that the identification is very inefficient. There are anyway methods to suppress this background:

(1) To produce a $\Delta$ the incident neutrino must have a momentum greater than 337 MeV/c, neglecting the nucleon Fermi motion. Furthermore the outgoing pion, to be detectable in water, must have a momentum greater than 159 MeV/c. In practice neutrinos with energy less than about 450 MeV cannot produce this kind of backgrounds. This is the main reason why CERN beta beam were initially proposed with $\gamma = 66$ [128, 129]: at those energies they are almost background free.

(2) The muon identification via the detection of a Michel electron is quite efficient in suppressing $\pi^-$ backgrounds, as the probability that a negative pion be absorbed before the completion of its decay chain is quite high.

There are two other aspects that need to be discussed about these backgrounds.

As first the kinematics of pion production is different from the kinematics of the neutrino charged-current interactions producing muons. This would suggest that an analysis of the emission angle with respect to the incoming neutrino energy could help in discriminating pions from muons. Unfortunately at neutrino energies below 1 GeV, Fermi motion makes this criteria inefficient.

The second important remark is about the momentum of the outgoing pions. The reaction is such that the momentum distribution peaks at the small values. This particular distribution is in most cases very different from signals coming from oscillated events, reducing the impact of pion backgrounds to the final sensitivities (see also Fig. 3.6).

The production rate of $\Delta$ resonant events increases quite fast with energy on the other hand increasing the neutrino energy makes both energy and angular selections more effective. In conclusion it is difficult if not impossible to make general conclusions about the effect of these backgrounds as a function of the $\gamma$ factor of parent ions without a detailed Monte Carlo simulation.

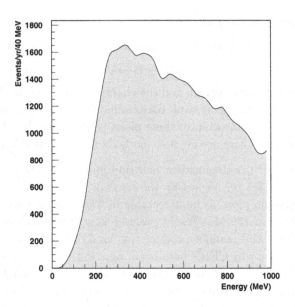

Fig. 3.1   Rate of atmospheric $\nu_\mu + \bar{\nu}_\mu$ interactions in MEMPHYS, integrated in one year.

### 3.3.1.3   *Backgrounds from atmospheric neutrinos*

This background source has an important impact on the beta beam design, so it will be discussed in some detail.

Atmospheric neutrinos are a continuous, isotropic flux of $\nu_e$, $\nu_\mu$, $\bar{\nu}_e$, $\bar{\nu}_\mu$ neutrinos. The spectrum of $\nu_\mu$ and $\bar{\nu}_\mu$ shown in Fig. 3.1 overlap the spectrum of oscillated signals, providing a copious source of backgrounds.

A first selection can be performed selecting events compatible with the neutrino beam direction. This selection cannot reduce the background very much because both the quasi-elastic kinematics and the Fermi motion conspire to generate a loose correlation between the outgoing lepton and the incoming neutrino. At the energies of the $\gamma = 100$ beta beam, the angular resolution is about 0.25 radians.

The only other handle is to keep the time in which beam neutrinos arrive to the detector very short, in other terms the duty cycle of the beta beam decay ring must be very short.

In [130] a full computation has been performed on the rate of atmospheric neutrino backgrounds in a beta beam experiment, taking into account the fluxes in the detector, the angular resolution and the efficiencies in detecting signal events. The results of this simulation indicate that only with a duty cycle of $10^{-2}$ does the atmospheric neutrino background rate go below the NC pion background rate.

As already stated, this in turn is the tightest constraint on the beta beam design derived from the optimization of the oscillation sensitivities.

At higher $\gamma$ the constraint on duty cycle relaxes because the atmospheric fluxes are less intense and the angular correlation tighter. This should help in loosening this constraint.

### 3.3.2 *Signals*

The neutrino flux in this setup is shown in Fig. 3.2. In this energy range almost all the neutrino charged-current interactions are quasi-elastic interactions (QE), a two-body configuration very favorable for a water Čerenkov detector.

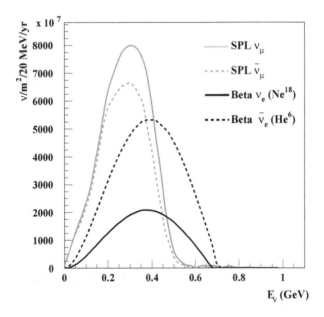

Fig. 3.2 Neutrino flux of $\beta$-Beam ($\gamma = 100$) together with the CERN-SPL super beam, at 130 km of distance.

Quasi-elastic interactions produce just one charged particle above the Čerenkov threshold, resulting in just one Čerenkov ring in the detector, a configuration where the event reconstruction results very efficient. Furthermore the two-body kinematics allow a precise reconstruction of the incident neutrino energy from the measured momentum of the outgoing lepton and the known direction of the incoming neutrino:

$$E_\nu^{\text{rec}} = \frac{1}{2} \frac{(M_p^2 - m_\mu^2) + 2E_\mu(M_n - V) - (M_n - V)^2}{-E_\mu + (M_n - V) + p_\mu \cos\theta_\mu} \quad (3.4)$$

where $M_p$, $M_n$, $m_\mu$, $E_\mu$, $p_\mu$, $\cos\theta_\mu$ $V$ are the proton, neutron, muon masses, muon energy, momentum and angle with respect to the incoming neutrino direction and the nuclear potential, set at 27 MeV, respectively. The performances of Super-Kamiokande in lepton momentum reconstruction are well-known, tested and documented, as well as the angular resolution.

The precision in measuring the neutrino energy is shown in Fig 3.3. The energy of non-quasi-elastic events reconstructed with this formula results underestimated, because of the different kinematics. This effect is hardly visible in the low energy bins, where the non-quasi-elastic event fraction is small.

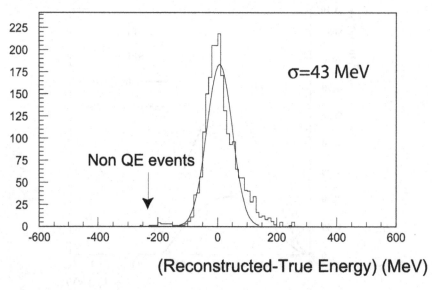

Fig. 3.3 Energy resolution for $\nu_e$ interactions in the 200–300 MeV energy range. The quantity displayed is the difference between the reconstructed and the true neutrino energy.

The non-gaussian features of energy reconstruction are taken into account by using migration matrices connecting true and reconstructed neutrino energy. They are computed by using NUANCE [131] and Super-

Table 3.1   Events in a 4400 kton/year exposure. $\nu_\mu(\overline{\nu}_\mu)$ CC events are computed assuming full oscillation, pion backgrounds are computed from $\nu_e(\overline{\nu}_e)$ CC+NC events. The three lines refer to interactions in the detector, selected after particle identification and selected after the detection of the Michel electron.

|  | Ne18 | | | He6 | | |
|---|---|---|---|---|---|---|
|  | $\nu_\mu$CC | $\pi^+$ | $\pi^-$ | $\overline{\nu}_\mu$CC | $\pi^+$ | $\pi^-$ |
| Interactions | 139181 | 863 | 561 | 107571 | 952 | 819 |
| Particle Id | 105923 | 209 | 123 | 83419 | 242 | 170 |
| Michel electron | 67888 | 103 | 6 | 67727 | 117 | 7 |

Kamiokande reconstruction algorithms (see [132], p. 139). 50 bins in true neutrino energy from 0 to 2 GeV are mapped onto 8 bins in reconstructed neutrino energy from 0.4 to 1.2 GeV. In total 8 migration matrices are used: for QE and non-QE events for each neutrino flavor $\nu_e$, $\overline{\nu}_e$, $\nu_\mu$, $\overline{\nu}_\mu$. Each matrix is normalized to take into account the single ring efficiency. The migration is consistently applied to signal and $\nu_e$ background events. Migration matrices are shown in Fig. 3.4.

Data reduction is shown in Fig. 3.5 for $^{18}$Ne events and detailed in Table 3.1 for $^{18}$Ne and $^6$He produced beam.

It is worth noting that the fraction of background events with respect to the fully oscillated sample, after the analysis selection, is about 0.2%, well below the $\sim 1\%$ characteristic of super beam experiments. Furthermore these backgrounds (Fig. 3.5) have a different spectral distribution from oscillated events, reducing their impact on oscillation analysis, as will be discussed in Section 3.4.1.

### 3.3.3   *Systematic errors*

Following the discussion of Section 1.4, one of the main reasons for the interest in beta beams in the context of ultimate long-baseline experiments, is the possibility of having a good control of systematic errors.

The first issue is the control of the beam: the beam spectrum at the far detector is fully described by the ion parents $\gamma$, their decay energy $E_\circ$, the experimental baseline $L$ and the number of ions circulating in the decay ring $N_\circ$. It seems feasible to keep these numbers under control in order to get an overall $\sim 1\%$ systematic error in the neutrino beam fluxes at the far detector. On the contrary super beam experiments need data from a dedicated hadroproduction experiment to precisely determine neutrino fluxes, having a $\sim 5\%$ precision as an ultimate goal.

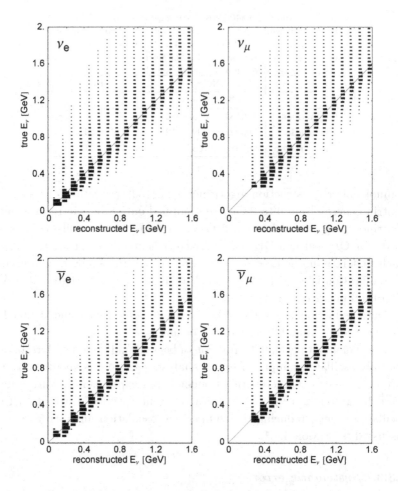

Fig. 3.4    Migration matrices for $\nu_e$, $\nu_\mu$, $\bar{\nu}_e$ and $\bar{\nu}_\mu$ events which pass the single ring cut and particle ID. The sum of quasi-elastic and non-quasi-elastic matrices is shown. The diagonal line denotes those points where reconstructed and true energy are the same.

Neutrino cross sections are necessary to estimate the interaction rates in the detector. In the energy range of the beta beam, neutrino cross section are poorly measured, with precisions of about 20%. The antineutrino cross sections are in an even worse situation, as well as the cross sections for the neutrino resonant production responsible for the pion backgrounds.

In the coming years the situation will certainly improve, thanks to the data that the SciBoone [133] and Miner$\nu$a [134] experiments, optimized

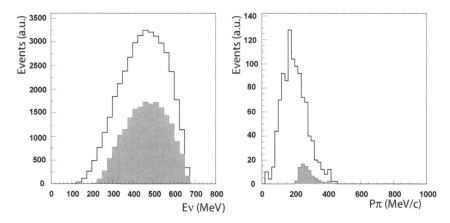

Fig. 3.5 Left: Event reduction for $^{18}$Ne oscillated events (left) and pion background, $\pi^+ + \pi^-$ (right).

to measure neutrino cross sections around 1 GeV, will collect, as well as the close detectors of T2K. Nevertheless it is difficult to predict systematic errors much better than 5 - 10% in the neutrino cross sections.

In this aspect however a close detector placed near the beta beam decay ring will have a good chance to significantly improve the experimental situation. The main reasons are:

- fluxes will be well known given the favorable condition of a beta beam experiment;
- signal data can be collected with negligible background, allowing a very clean data sample for the cross section measurement;
- the resonant production of charged pions can also be measured with very little background, since they produce charged pions in the final state that can efficiently separated from the electron produced by the beam charged current interactions;
- the $\gamma$ of the parent ions can be varied in the accelerator complex, allowing for a full scan of neutrino energies.

Under these conditions, a precision of a few percent in signal and background cross section measurements seems to be a realistic goal.

The weak point of cross section measurements in a beta beam close detector is derived from its own purity. It has been shown in [64] that a very important factor in LCPV searches is the determination of the $\nu_e/\nu_\mu$

and $\overline{\nu}_e/\overline{\nu}_\mu$ cross section ratios. No $\nu_\mu$ or $\overline{\nu}_\mu$ are present at the close detector site, so these two cross section ratios cannot be measured in a beta beam close detector.

The possible way outs are:

- Theoretical models about neutrino cross section can be improved, allowing for a stringent definition of the cross section ratios. First principles would say that $\nu_e$ and $\nu_\mu$ charged-current interactions differ only for the lepton mass, and that this correction factor (sizable at beta beam energies) can be precisely computed. Nuclear effects could anyway spoil these predictions, introducing systematic effects around 2 - 5%. A focused effort in this direction is lacking in the literature and it should be energetically pursued in case of a beta beam startup.
- A conventional neutrino beam, generating $\nu_\mu$ and $\overline{\nu}_\mu$ interactions, could be built ad hoc for this measure in the beta beam close detector. It does not necessarily have to be a high intensity beam, but its design should be tailored to achieve good precision in the cross section measurement.

## 3.4   Oscillation Analysis

In the following the CERN-Fréjus beta beam capabilities in measuring oscillation processes will be discussed.

The physics analysis is performed with the GLoBES open source software [135], which provides a convenient tool to simulate long-baseline experiments and compare different facilities in a unified framework. The experiment definition (AEDL) files for the beta beam simulation with GLoBES are available in [135].

Most of the results shown in the following are taken from [76].

Sensitivities in discovering non-zero values of $\theta_{13}$ will be discussed in Section 3.4.1, followed by sensitivities in discovering leptonic CP violation in Section 3.4.2. In these analyses parameter degeneracies and correlations are fully taken into account and in general all oscillation parameters are varied in the fit. These sections will be followed by a discussion on the limitation of the CERN-Fréjus setup (Section 3.4.3) namely the lack of sensitivity in measuring the atmospheric parameters, in measuring $\text{sign}(\Delta m_{23}^2)$, in unambiguously determining $\theta_{13}$ and $\delta_{\text{CP}}$ in case of signal and in solving all the degenerate solutions.

However these limitations can be overcome if the beta beam signals are combined with the atmospheric neutrino signals that can be collected

Table 3.2 Summary of default parameters used for the simulation of the beta beam experiment.

| | |
|---|---|
| Detector mass | 440 kt |
| Baseline | 130 km |
| Running time $(\nu + \bar{\nu})$ | 5 + 5 yr |
| Beam intensity | $5.8\,(2.2) \cdot 10^{18}$ He (Ne) dcys/yr |
| Systematics on signal | 2% |
| Systematics on backgr. | 2% |

"for free" in the MEMPHYS detector (Section 3.5). This will allow for a considerable sensitivity in measuring sign($\Delta m_{23}^2$) (Section 3.5.1) and in breaking degeneracies (Section 3.5.2) providing an unambiguous determination of $\theta_{13}$ and $\delta_{CP}$ in case of signal in a given subset of the parameter space.

Finally the synergies of a combined data taking with neutrinos from the SPL super beam will be illustrated in Section 1.3.4.4, followed by an overall comparison with the sensitivities that can be achieved by some super beam projects (Section 3.7).

The default parameters used in the following are listed in Table 3.2.

To simulate the "data" the following set of "true values" for the oscillation parameters is adopted:

$$\Delta m_{31}^2 = +2.4 \times 10^{-3} \text{ eV}^2 \,, \qquad \sin^2 \theta_{23} = 0.5 \,,$$
$$\Delta m_{21}^2 = 7.9 \times 10^{-5} \text{ eV}^2 \,, \qquad \sin^2 \theta_{12} = 0.3 \,, \qquad (3.5)$$

a prior knowledge of these values is included with an accuracy of 10% for $\theta_{12}$, $\theta_{23}$, $\Delta m_{31}^2$, and 4% for $\Delta m_{21}^2$ at $1\sigma$. These values and accuracies are motivated by recent global fits to neutrino oscillation data [21, 22], and they are always used except where explicitly stated otherwise.

### 3.4.1 $\theta_{13}$ searches

Non-zero values of $\theta_{13}$ are looked for by exploiting $\nu_e \rightarrow \nu_\mu$ transitions. Following Eq. (1.4), $\nu_e \rightarrow \nu_\mu$ transitions can occur even for null values of $\theta_{13}$, thanks to the contribution of the "solar" terms.

The detection of a transition rate in excess of the solar terms is the main signature of non-zero values of $\theta_{13}$. The sensitivity to $\theta_{13}$ (or to $\sin^2 2\theta_{13}$) is defined as the smallest value of $\theta_{13}$ ($\sin^2 2\theta_{13}$) which can be distinguished from $\theta_{13} = 0$. The final sensitivity to these transitions is characterized by a small number of signal events, to be disentangled from backgrounds.

Table 3.3   Number of events for appearance and disappearance signals and backgrounds for the beta beam experiment as defined in Tab. 3.2. For the appearance signals the event numbers are given for several values of $\sin^2 2\theta_{13}$ and $\delta_{CP} = 0$ and $\pi/2$. The background as well as the disappearance event numbers correspond to $\theta_{13} = 0$. For the other oscillation parameters the values of Eq. (3.5) are used. From [76].

|  | $\delta_{CP} = 0$ | $\delta_{CP} = \pi/2$ |
|---|---|---|
| appearance $\nu$ |  |  |
| background |  | 143 |
| $\sin^2 2\theta_{13} = 0$ |  | 28 |
| $\sin^2 2\theta_{13} = 10^{-3}$ | 76 | 88 |
| $\sin^2 2\theta_{13} = 10^{-2}$ | 326 | 365 |
| appearance $\bar{\nu}$ |  |  |
| background |  | 157 |
| $\sin^2 2\theta_{13} = 0$ |  | 31 |
| $\sin^2 2\theta_{13} = 10^{-3}$ | 83 | 12 |
| $\sin^2 2\theta_{13} = 10^{-2}$ | 351 | 126 |
| disappearance $\nu$ |  | 100315 |
| disappearance $\bar{\nu}$ |  | 84125 |

Under this aspect a beta beam setup appears to be very attractive, having a very small number of backgrounds with an energy spectrum significantly different from oscillation signals, as illustrated in Table 3.3 and Fig. 3.6.

Systematic errors do not play a major role in these searches, where the important features are a high rate of neutrino events and a low rate of background events.

From Eq. (1.4) and Fig. 1.2, it can be also noted that the value of $\delta_{CP}$ can greatly influence the sensitivity to $\theta_{13}$. This suggests a combined run with neutrinos and antineutrinos (where the $\delta_{CP}$ effect is opposite sign) to reduce such an influence.

The procedure to determine the $\theta_{13}$ discovery potential is the following: data are simulated for a finite true value of $\sin^2 2\theta_{13}$ and a given true value of $\delta_{CP}$. If the $\Delta\chi^2$ of the fit to these data with $\theta_{13} = 0$ is larger than 9 the corresponding true value of $\theta_{13}$ "is discovered at $3\sigma$". In other words, the $3\sigma$ discovery limit as a function of the true $\delta_{CP}$ is given by the true value of $\sin^2 2\theta_{13}$ for which $\Delta\chi^2(\theta_{13} = 0) = 9$. In the fitting process $\Delta\chi^2$

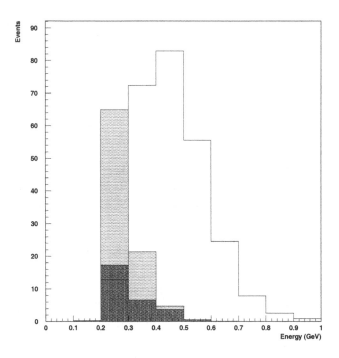

Fig. 3.6 Event rates for $\sin^2 2\theta_{13} = 0.01$, $\delta_{CP} = 0$, for a 4x440 kt/year exposure of $^{18}$Ne beam. Solid lines are oscillation signals, dashed lines backgrounds from NC events, dotted lines backgrounds from atmospheric neutrino interactions.

is minimized with respect to $\theta_{12}$, $\theta_{23}$, $\Delta m_{12}^2$, and $\Delta m_{31}^2$, and in general one has to test also for degenerate solutions in $\text{sign}(\Delta m_{31}^2)$ and the octant of $\theta_{23}$.

As anticipated in Section 1.2.2, degenerate solutions do not influence $\theta_{13}$ sensitivity very much, for the simple fact that for very small $\theta_{13}$ these degenerate solutions disappear.

The discovery limits are shown in Fig. 3.7.

One can note that a guaranteed discovery reach of $\sin^2 2\theta_{13} \simeq 5 \times 10^{-3}$ is obtained, irrespective of the actual value of $\delta_{CP}$. However, for certain values of $\delta_{CP}$ the sensitivity is significantly improved. A sensitivity below $\sin^2 2\theta_{13} = 4 \times 10^{-4}$ is reached for 80% of all possible values of $\delta_{CP}$. The beta beam performance depends crucially on the neutrino flux intensity, as can be seen from the dashed curves in Fig. 3.7, which has been obtained by reducing the number of ion decays/yr by a factor of two with respect to our standard values given in Tab. 3.2. In this case the sensitivity decreases significantly.

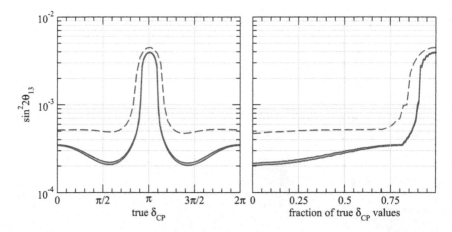

Fig. 3.7  $3\sigma$ discovery sensitivity to $\sin^2 2\theta_{13}$ for beta beam, as a function of the true value of $\delta_{CP}$ (left panel) and as a function of the fraction of all possible values of $\delta_{CP}$ (right panel). The running time is $(5\nu + 5\bar\nu)$ yrs. The width of the bands corresponds to values for systematical errors between 2% and 5%. The dashed curves show the sensitivity of the beta beam when the number of ion decays/yr is reduced by a factor of two with respect to the values given in Table 3.2.

The peak of the sensitivity curves around $\delta_{CP} \approx \pi$ appears due to the interplay of neutrino and antineutrino data. The particular shape of the sensitivity curves emerges from the relative location of the corresponding curves for neutrino and antineutrino data, which is controlled by the $L/E_\nu$ value where the experiment is operated and the value of $|\Delta m^2_{31}|$. The fact that the peak is so pronounced follows from the fact that in the CERN-Fréjus configuration $L/E_\nu$ is relatively small.

In Fig. 3.7 we illustrate also the effect of systematical errors on the $\theta_{13}$ discovery reach. The lower boundary of the band for each experiment corresponds to a systematical error of 2%, whereas the upper boundary is obtained for 5%. These errors include the (uncorrelated) normalization uncertainties on the signal as well as on the background, where the crucial uncertainty is the error on the background. The beta beam is basically not affected by these errors, since the background has a rather different spectral shape (strongly peaked at low energies) than the signal. This is in contrast with super beam experiments, as discussed in [76].

Let us remark that the $\theta_{13}$ sensitivities are practically not affected by the $\text{sign}(\Delta m^2_{31})$-degeneracy. This is easy to understand, since the data is fitted with $\theta_{13} = 0$, and in this case both mass hierarchies lead to very

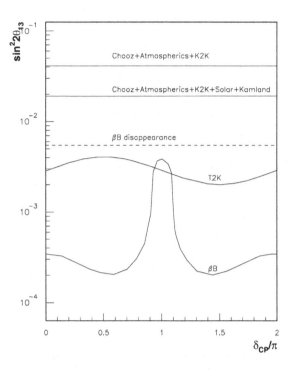

Fig. 3.8   The same sensitivity curve as Fig. 3.7 (2% systematic errors), compared with the present world limits on $\theta_{13}$ [22] (solid lines), the $3\sigma$ T2K sensitivity computed for a 10 year neutrino run, and the CFBB sensitivity for the disappearance channel, computed for 1% systematic errors.

similar event rates. If the inverted hierarchy is used as the true hierarchy, the peak in the discovery limit visible in the left panel of Fig. 3.7 around $\delta_{\rm CP} \sim \pi$ moves to $\delta_{\rm CP} \sim 0$. However, the characteristic shape of the curves, and in particular, the sensitivity as a function of the $\delta_{\rm CP}$-fraction shown in the right panel, are hardly affected by the sign of the true $\Delta m^2_{31}$. In case of a non-maximal value of $\theta_{23}$ the octant-degeneracy has a minor impact on the $\theta_{13}$ discovery potential.

Also $\nu_e \to \nu_e$ transitions contribute to the $\theta_{13}$ sensitivity. They are however marginal if the overall systematic error is around 2% (as a comparison reactor experiments plan to reach systematic errors of about 0.2% in $\bar{\nu}_e$ disappearance just to reach sensitivities of $\sin^2 2\theta_{13} \simeq 0.01$). As computed in [75], the CFBB experiment could reach sensitivities of $\sin^2 2\theta_{13} \leq 0.02$ (90% CL) to $\nu_e$ disappearance. Such values, compared to the sensitivity of Fig. 3.7, are clearly marginal.

### 3.4.2 *Leptonic CP violation searches*

In case a finite value of $\theta_{13}$ is established it is important to quantitatively assess the discovery potential for leptonic CP violation (LCPV). The CP symmetry is violated if the complex phase $\delta_{CP}$ is different from 0 and $\pi$. Therefore, LCPV is discovered if these values for $\delta_{CP}$ can be excluded.

Leptonic CP violation searches are performed by comparing event rates and spectra in neutrino and antineutrino runs, as discussed in Section 1.2.1. There are two possible regimes in this search: for relatively large values of $\theta_{13}$ (say $\sin^2 2\theta_{13} > 0.01$), signal event rates are rather large, while the asymmetry between neutrino and antineutrino rates is relatively small. In this condition background rates are not that important and the dominant factor is systematic errors.

For relatively small values of $\theta_{13}$, signal rates are small, while the asymmetry is large. Under this condition systematic errors are less important, while background rates become an issue.

The LCPV sensitivity curves are calculated by scanning the true values of $\sin^2 2\theta_{13}$ and $\delta_{CP}$. Then these data are fitted with the CP conserving values $\delta_{CP} = 0$ and $\delta_{CP} = \pi$, where all parameters except $\delta_{CP}$ are varied and the sign and octant degeneracies are taken into account. If no fit with $\Delta\chi^2 < 9$ is found, CP conserving values of $\delta_{CP}$ can be excluded at $3\sigma$ for the chosen values of $\delta_{CP}^{\text{true}}$ and $\sin^2 2\theta_{13}^{\text{true}}$.

The LCPV discovery potential is shown in Fig. 3.9. For systematical errors of 2% maximal LCPV (for $\delta_{CP}^{\text{true}} = \pi/2, 3\pi/2$) can be discovered at $3\sigma$ down to $\sin^2 2\theta_{13} \simeq 2 \times 10^{-4}$. This number is increased by a factor of 3 if the fluxes are reduced to half of the nominal values. The best sensitivity to LCPV is obtained around $\sin^2 2\theta_{13} \sim 10^{-2}$. For this value LCPV can be established for 78% of all values of $\delta_{CP}$ (again for systematics of 2%). The widths of the bands in Fig. 3.9 correspond to different values of the systematical errors; it turns out that the most relevant uncertainty is the background normalization.

The impact of systematics is very small for the beta beam. The reason for this is that the spectral shape of the background in the beta beam (from pions and atmospheric neutrinos) is very different from the signal, and therefore they can be disentangled by the fit of the energy spectrum.

One finds that systematical errors dominate ($\sigma_{\text{bkgr}}\sqrt{B} > 1$) if $\sigma_{\text{bkgr}} \gtrsim$ 6%. In the right panel of Fig. 3.10 we show the sensitivity to maximal LCPV (as defined in the figure caption) as a function of $\sigma_{\text{bkgr}}$. Indeed, the

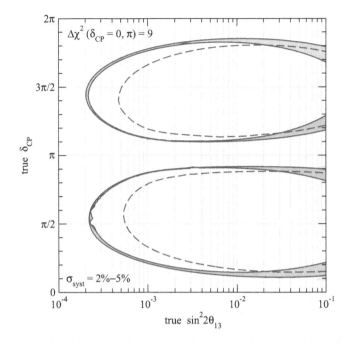

Fig. 3.9 LCPV discovery potential: for parameter values inside the ellipse-shaped curves CP conserving values of $\delta_{CP}$ can be excluded at $3\sigma$ ($\Delta\chi^2 > 9$). The running time is $(5\nu + 5\bar{\nu})$ yrs. The width of the bands corresponds to values for the systematical errors from 2% to 5%. The dashed curves show the sensitivity when the number of ion decays/yr are reduced by a factor of two with respect to the values given in Table 3.2 for 2% systematics.

worsening of the sensitivity due to systematics occurs roughly at the values of $\sigma_{\text{bkgr}}$ as estimated above.

The left panel of Fig. 3.10 shows the sensitivity to maximal LCPV as a function of the exposure for values of $\sigma_{\text{bkgr}}$ from 2% to 5%. One can observe clearly that for the standard exposure of 4400 kt yrs sensitivity is rather stable with respect to systematics and for the standard exposure it is still statistics dominated.

Finally, Fig. 3.11 illustrates the impact of degeneracies, as well as the true value of sign($\Delta m_{23}^2$) and the true $\theta_{23}$-octant on the LCPV sensitivity. Curves of different strokes correspond to the four different choices for sign($\Delta m_{23}^2$) and $\theta_{23}$-octant of the true parameters. For the solid curves the simulated data for each choice of true sign($\Delta m_{23}^2$) and $\theta_{23}$-octant are fitted by taking into account all four degenerate solutions, i.e., also for the fit all four combinations of sign($\Delta m_{23}^2$) and $\theta_{23}$-octant are used.

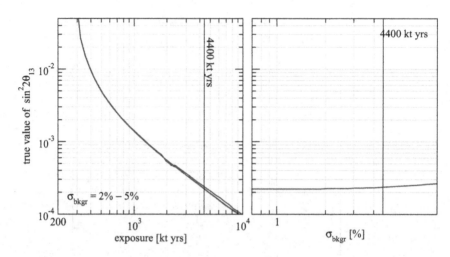

Fig. 3.10 Impact of total exposure and systematical errors on the LCPV discovery potential. We show the smallest true value of $\sin^2 2\theta_{13}$ for which $\delta_{CP} = \pi/2$ can be distinguished from $\delta_{CP} = 0$ or $\delta_{CP} = \pi$ at $3\sigma$ ($\Delta\chi^2 > 9$) as a function of the exposure in kt yrs (left) and as a function of the systematical error on the background $\sigma_{bkgr}$ (right). The widths of the curves in the left panel corresponds to values of $\sigma_{bkgr}$ from 2% to 5%. The thin solid curves in the left panel correspond to no systematical errors. The right plot is calculated for the standard exposure of 4400 kt yrs. No systematical error on the signal has been assumed.

One observes from the figure that the true hierarchy and octant have a rather small impact on the LCPV sensitivity. In particular the sensitivity to maximal LCPV is completely independent. The main effect of changing the true hierarchy is to exchange the behavior between $0 < \delta_{CP} < 180°$ and $180° < \delta_{CP} < 360°$. For $\sin^2 2\theta_{13} \lesssim 10^{-2}$ the sensitivity gets slightly worse if $\theta_{23}^{\ true} > \pi/4$ compared to $\theta_{23}^{\ true} < \pi/4$.

To appreciate the effect of degeneracies in LCPV sensitivity, the dashed curves in Fig. 3.11 are computed without taking into account the degeneracies, i.e., for each choice of true $\text{sign}(\Delta m_{23}^2)$ and $\theta_{23}$-octant the data are fitted only with this particular choice. The effect of the degeneracies becomes visible for large values of $\theta_{13}$. Note that this is just the region where they can be reduced by a combined analysis with atmospheric neutrinos (see Section 3.5).

### 3.4.3 *Searches that cannot be done in this configuration*

The physics potential of the CERN-Fréjus project is quite impressive as far as sensitivity to $\theta_{13}$ and LCPV is concerned.

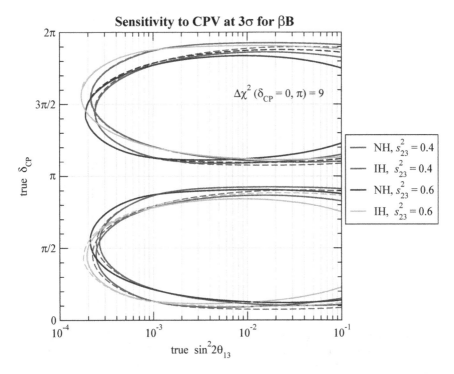

Fig. 3.11 Impact of degeneracies on the LCPV discovery potential. Sensitivity to CV at $3\sigma$ ($\Delta\chi^2 > 9$) is computed for 4 different combinations of the true values of the hierarchy (NH or IH) and $\theta_{23}$ ($\sin^2\theta_{23} = 0.4$ or 0.6). Dashed curves are computed neglecting degeneracies in the fit. The running time is ($5\nu + 5\bar{\nu}$) yrs. From [76].

There are, anyway, measurements of the oscillation parameters that are not within the reach of this project, if the analysis is limited to the beta beam data. Most of these limits are overcome if the beta beam data are combined with atmospheric neutrinos, so the following Section 3.5 is the natural complement to this discussion.

First the precision with which the atmospheric parameters $\theta_{23}$ and $\Delta m^2_{23}$ are known cannot be improved by a beta beam experiment, for the simple fact that these parameters can be precisely measured only with $\nu_\mu$ disappearance data and a beta beam does not provide any $\nu_\mu$ beam. This has little influence on $\theta_{13}$ and LCPV, also because the precision with which the T2K experiment will measure the atmospheric parameters is already adequate for the needs of these searches.

The baseline of 130 km is insufficient to produce sizable matter effects in the CERN-Fréjus experiment, resulting in a very poor sensitivity to the

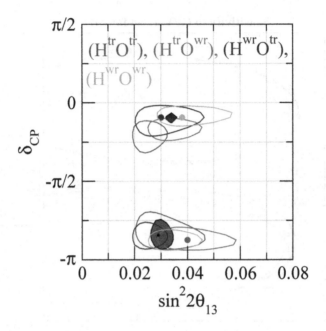

Fig. 3.12   Allowed regions in $\sin^2 2\theta_{13}$ and $\delta_{CP}$ for beta beam data alone (contour lines) and LBL+ATM data combined (shaded regions). $\mathrm{H^{tr/wr}(O^{tr/wr})}$ refers to solutions with the true/wrong mass hierarchy (octant of $\theta_{23}$). The true parameter values are $\delta_{CP} = -0.85\pi$, $\sin^2 2\theta_{13} = 0.03$, $\sin^2 \theta_{23} = 0.6$, and the values from Eq. (3.5) for the other parameters. The running time is $(5\nu + 5\bar{\nu})$ yrs. From [76].

value of $\mathrm{sign}(\Delta m_{23}^2)$. On the other hand this is highly beneficial for LCPV searches, because matter effects are one of the main sources of background and distortions for these analyses.

The effect of degeneracies is illustrated for a particular choice of the parameters in Fig. 3.12 (see also [76]). Assuming the true parameter values $\delta_{CP} = -0.85\pi$, $\sin^2 2\theta_{13} = 0.03$, $\sin^2 \theta_{23} = 0.6$, the allowed regions in the plane of $\sin^2 2\theta_{13}$ and $\delta_{CP}$ are shown, taking into account the solutions with the wrong hierarchy and the wrong octant of $\theta_{23}$.

One observes from Fig. 3.12 that in this case the $(\delta_{CP}, \theta_{13})$-degeneracy cannot be resolved and one has to deal with eight distinct solutions. One reason for this is the absence of precise information on $|\Delta m_{31}^2|$ and $\sin^2 2\theta_{23}$ which is provided by the $\nu_\mu$ disappearance in super beam experiments but is not available from the beta beam. If external information on these parameters at the level of 3% is included (this is the case of the T2K experiment), the allowed regions in Fig. 3.12 are significantly reduced. However, still all eight solutions are present, which indicates that the CFBB spectral information is not efficient enough to resolve the $(\delta_{CP}, \theta_{13})$-degeneracy, and in this case only the inclusion of atmospheric neutrino data allows a nearly complete resolution of the degeneracies.

An important observation from Fig. 3.12 is that degeneracies have only a very small impact on the CP violation discovery, in the sense that if the true solution is CP violating the fake solutions are also located at CP violating values of $\delta_{CP}$. Indeed, since matter effects are very small for the relatively short baselines in the experiments under consideration, the $\text{sign}(\Delta m_{31}^2)$-degenerate solution is located within good approximation at $\delta'_{CP} \approx \pi - \delta_{CP}$ [33]. Therefore, although degeneracies strongly affect the determination of $\theta_{13}$ and $\delta_{CP}$ they have only a small impact on the CP violation discovery potential. Furthermore, as is clear from Fig. 3.12 the $\text{sign}(\Delta m_{31}^2)$ degeneracy has practically no effect on the $\theta_{13}$ measurement, whereas the octant degeneracy has very little impact on the determination of $\delta_{CP}$.

## 3.5 Combined Analyses with the Atmospheric Neutrinos

Beta beam and atmospheric neutrino data are a truly synergic combination, in that together the two samples provide more information than expected just from statistics.

It has just been discussed that beta beam has very limited capabilities in measuring $\text{sign}(\Delta m_{23}^2)$ and resolving degeneracies on the other hand atmospheric neutrinos, even if measured with large statistics, cannot measure $\text{sign}(\Delta m_{23}^2)$ in the absence of a measured value of $\theta_{13}$, precisely what beta beam measures at best.

The power of a combination of LBL experiments based on megaton scale water Čerenkov detectors with data from atmospheric neutrinos (ATM) has been pointed out in [37]. Atmospheric neutrinos are sensitive to the neutrino mass hierarchy if $\theta_{13}$ is sufficiently large due to Earth matter effects, mainly in multi-GeV $e$-like events [136–138]. Moreover, sub-GeV $e$-like events provide sensitivity to the octant of $\theta_{23}$ [139–141] due to oscillations with $\Delta m_{21}^2$ (see also reference [142] for a discussion of atmospheric neutrinos in the context of Hyper-Kamiokande).

A detailed computation of the beta beam+ATM analysis has been performed in [76], where the ATM analysis was tailored to the characteristics of the MEMPHYS detector, whose bigger dimensions with respect to Super-

Kamiokande allow for the containment of events of higher energy. Also multi-ring events, defined as fully contained charged-current events which are not tagged as single-ring, are included in the analysis.

In this analysis three different kinds of experimental uncertainties are included: flux uncertainties: total normalization (20%), tilt factor (5%), zenith angle (5%), $\nu/\bar{\nu}$ ratio (5%), and $\mu/e$ ratio (5%); cross-section uncertainties: total normalization (15%) and $\mu/e$ ratio (1%) for each type of charged-current interaction (quasi-elastic, one-pion production, and deep-inelastic scattering), and total normalization (15%) for the neutral-current contributions; systematic uncertainties: same as in previous analyses, details are given in the Appendix of [143]. An independent normalization uncertainties (20%) for $e$-like and $\mu$-like multi-ring events is added to these terms.

### 3.5.1  *Mass hierarchy*

The combination of ATM+beta beam data leads to a non-trivial sensitivity to the neutrino mass hierarchy, i.e. to the sign of $\Delta m_{31}^2$ as shown in Fig. 3.13. For beta beam data alone (dashed curves) there is practically no sensitivity in the CERN–MEMPHYS experiment (because of the very small matter effects due to the relatively short baseline). However, by including data from atmospheric neutrinos (solid curves) the mass hierarchy can be identified at $2\sigma$ CL provided $\sin^2 2\theta_{13} \gtrsim 0.02 - 0.03$. Fig. 3.13 is computed with a true value of $\theta_{23} = \pi/4$. Generically the hierarchy sensitivity increases with increasing $\theta_{23}$, see [37] for a detailed discussion.

### 3.5.2  *Degeneracy breaking*

The effect of the atmospheric data in breaking degeneracies has been shown in Section 3.4.3. It is worth adding that atmospheric neutrinos can help in breaking the octant degeneracy. In this case there is no synergy with beta beam data because what would be needed is information on the value of $|\Delta m_{23}^2|$. Fig. 3.14 shows the potential of ATM data to exclude the octant degenerate solution.

## 3.6  Combined Analyses with the SPL Super Beam

Soon after the first proposal of beta beams [85] it was realized that neutrinos created by the SPL could be fired to the same detector [75].

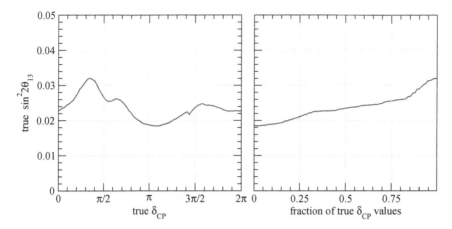

Fig. 3.13 Sensitivity to the mass hierarchy at $2\sigma$ ($\Delta\chi^2 = 4$) as a function of the true values of $\sin^2 2\theta_{13}$ and $\delta_{CP}$ (left), and the fraction of true values of $\delta_{CP}$ (right). The solid curves are the sensitivities from the combination of long-baseline and atmospheric neutrino data, the dashed curves correspond to beta beam data only. The running time is $(5\nu + 5\bar{\nu})$ yrs.

The injector of a beta-beam complex must be a 1 - 3 GeV Linac, precisely the energy of the SPL. Furthermore radioactive ion production requires at most 0.1 MW, while SPL could deliver up to 4 MW of power.

Under these circumstances a very intense super beam, already discussed in Section 1.3.4.4, can run together with a beta beam. The typical energy of a neutrino beam created by the SPL can nicely match the energy of a $\gamma = 100$ beta beam (see Fig. 3.2) so the two neutrino beams can share the same baseline, thus the same detector.

The combination of a super beam with a beta beam in the same experiment can provide an experimental environment with very unique characteristics:

- The two beams can be used to separately study CP channels like $\nu_\mu \to \nu_e$ vs $\bar{\nu}_\mu \to \bar{\nu}_e$ and $\nu_e \to \nu_\mu$ vs $\bar{\nu}_e \to \bar{\nu}_\mu$.
- They can be mixed to study T transitions like $\nu_\mu \to \nu_e$ vs $\nu_e \to \nu_\mu$ and $\bar{\nu}_\mu \to \bar{\nu}_e$ vs $\bar{\nu}_e \to \bar{\nu}_\mu$.
- The can be mixed to study CPT transitions like $\nu_\mu \to \nu_e$ vs $\bar{\nu}_e \to \bar{\nu}_\mu$ and $\nu_e \to \nu_\mu$ vs $\bar{\nu}_\mu \to \bar{\nu}_e$.

The addition of a super beam to a beta beam could also complement some of the weak points of the beta beam, namely the lack of sensitivity to the atmospheric parameters $\theta_{23}$ and $\Delta m^2_{23}$ and the lack of $\nu_\mu$ events in

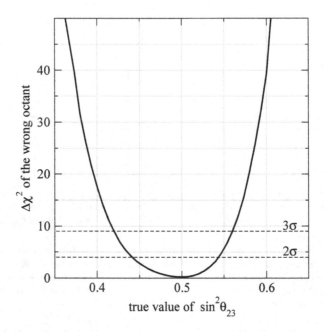

Fig. 3.14 $\Delta\chi^2$ of the solution with the wrong octant of $\theta_{23}$ as a function of the true value of $\sin^2\theta_{23}$. Computed for a true value of $\theta_{13} = 0$, and a running time of 10 yrs.

the close detector, useful for calibrating beta beam signal efficiency and measuring the $\nu_e/\nu_\mu$ cross section ratio.

The only other experimental setup that could match such a broad range of transitions is a neutrino factory, where however the different transitions would probably need different specialized detectors and some channels would require detector performances well beyond the present state of the art (for instance $\nu_\mu \to \nu_e$ and $\bar{\nu}_\mu \to \bar{\nu}_e$ transitions in a Neutrino Factory require a detector capable of measuring electron charges for events up to 5 GeV with a charge separation of $10^3$ or better).

In an SPL super beam+beta beam experiment all the channels would instead be measured in the same detector with small background rates. This is highly beneficial for systematic errors and would provide redundancy in the oscillation signals, a feature that should not be underestimated in an experimental field that today is completely unexplored.

From the strict point of LCPV searches and degeneracy breaking, SPL+beta beam is not a synergic combination, as noted in [40], since the two beams have identical baselines and similar neutrino energies. This means that the combination of the two beams does not add anything significant other than a simple increase of statistics. Nevertheless the combination of the two significantly improves overall performances, since they are still limited by statistics and not by systematics (even assuming 5% systematic errors). This is shown in Figs. 3.15, 3.16, 3.17, taken from [76]. These plots contain information pertinent also to next Section 3.7.

As far as is concerned sign($\Delta m_{23}^2$) sensitivity the combination SPL+beta beam provides rather good sensitivity even without atmospheric data, and in this aspect the two beams are synergic. Because of the rather short baseline the matter effect is too small to distinguish between normal and inverted hierarchy given only neutrino and antineutrino information in one channel. However, the tiny matter effect suffices to move the hierarchy degenerate solution to slightly different locations in the ($\sin^2 2\theta_{13}$, $\delta_{CP}$) plane for the $\overset{(-)}{\nu}_e \to \overset{(-)}{\nu}_\mu$ (beta beam) and $\overset{(-)}{\nu}_\mu \to \overset{(-)}{\nu}_e$ (SPL) channels (compare Fig. 3.12). Hence, if all four CP and T conjugate channels are available already the small matter effect picked up along the 130 km CERN–MEMPHYS distance provides sensitivity to the mass hierarchy for $\sin^2 2\theta_{13} \gtrsim 0.03$, or $\sin^2 2\theta_{13} \gtrsim 0.015$ if atmospheric neutrino data is also included. The effect is shown in Fig. 3.17.

## 3.7 Comparison with Other Super Beam Experiments

In this section the beta beam performances are compared with those of two super beam projects: T2HK (presented in Section 1.3.4.2) and the SPL super beam (see Section 1.3.4.4).

A comparison with a larger set of proposed facilities will be discussed in Section 4.6.

The discussion is restricted to T2HK and the SPL super beam for the main reason that these projects share the same detector technology: a megaton scale water Čerenkov detector. So detector performances are the same, and well defined given the experience accumulated in the Super-Kamiokande and K2K experiments.

In a certain sense this discussion addresses the following question: which is the best beam that can be fired to a megaton scale water Čerenkov detector?

Results of this comparison are shown in Figs. 3.15, 3.16, 3.17. It can be seen that CFBB outperforms the two super beams as far as is concerned $\theta_{13}$ and LCPV discovery potential while T2HK has a better sensitivity to sign($\Delta m_{23}^2$), thanks to the longer baseline. It should also be noted that

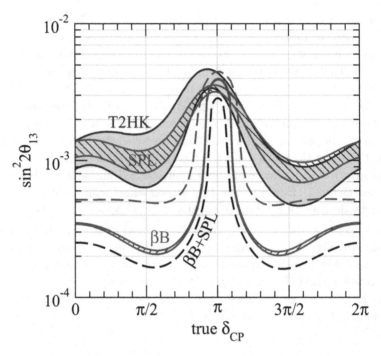

Fig. 3.15 $3\sigma$ discovery sensitivity to $\sin^2 2\theta_{13}$ for beta beam, SPL super beam, and T2HK as functions of the true value of $\delta_{CP}$. The running time is $(5\nu + 5\bar{\nu})$ yrs for beta beam and $(2\nu + 8\bar{\nu})$ yrs for the super beams. The width of the bands corresponds to values for systematical errors between 2% and 5%. The black curves correspond to the combination of beta beam and SPL with 10 yrs of total data taking each for a systematical error of 2%, and the dashed curves show the sensitivity of the beta beam when the number of ion decays/yr are reduced by a factor of two with respect to the values given in Table 3.2. From [76].

beta beam performances are less sensitive to a change from 2% to 5% of the systematic errors, while T2HK has the biggest sensitivity variation in changing the systematic errors. This is due to the absolute number of background events in the different setups. Beta beam performances are very much dependent on the neutrino fluxes, and indeed the main factor in increasing beta beam performances is the value of neutrino fluxes. Finally the combination of beta beam+SPL neutrino beams would produce excellent physics potential, especially as far as is concerned LCPV sensitivity.

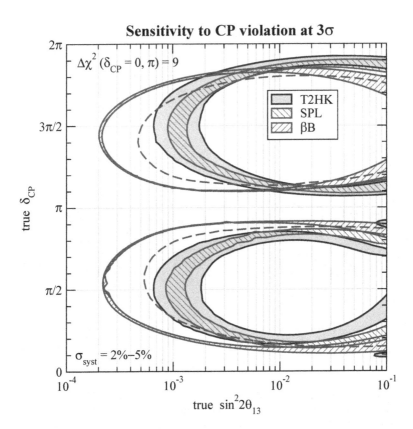

Fig. 3.16 LCPV discovery potential, for parameter values inside the ellipse-shaped curves CP conserving values of $\delta_{CP}$ can be excluded at $3\sigma$ ($\Delta\chi^2 > 9$). Computed with identical assumptions as Fig. 3.15 From [76].

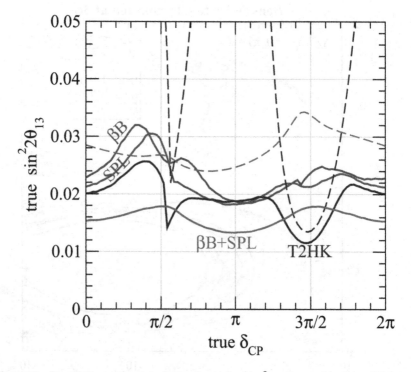

Fig. 3.17   Sensitivity to the mass hierarchy at $2\sigma$ ($\Delta\chi^2 = 4$) as a function of the true values of $\sin^2 2\theta_{13}$ and $\delta_{CP}$. The solid curves are the sensitivities from the combination of long-baseline and atmospheric neutrino data, the dashed curves correspond to long-baseline data only. The running time is ($5\nu + 5\bar{\nu}$) yrs for the beta beam and ($2\nu + 8\bar{\nu}$) yrs for the super beams. From [76].

# Chapter 4

# Physics Potential of Other Beta Beam Settings

## 4.1 Introduction

The CERN-Fréjus beta beam setup is based on existing machines and possible upgrades of underground laboratories, as already discussed in Section 3.1; it is designed to be a realistic configuration, not necessarily the optimal one.

This leaves room for studies devoted to optimal beta beam configurations, where the potential of different setups is studied with fewer constraints about their final implementation. These "green field" scenarios match the approach of neutrino factories, that have studied from the beginning optimal setups for the production of neutrino beams. Perhaps it is not surprising that eventually optimal beta beams and optimal neutrino factories have rather similar performances.

The first option studied in the literature has been high energy beta beams, which will be presented in Section 4.2. These configurations consider the same setup as the CFBB except for the final accelerator, where a machine capable of accelerating protons up to 1 TeV (more than twice the energy of the SPS) is taken into account. This reflects also a more demanding decay ring configuration.

Following the chronology of literature publications, Section 4.3 will discuss monochromatic beta beams. These beams are based on electron capture processes of radioactive ions, rather than on their beta decays, producing monochromatic neutrino beams. This is an extremely interesting setup, since the neutrino detector has only to guarantee a correct particle identification, being the neutrino energy known at the source. As already discussed in Section 2.7.5 the main limitations of these setups are the technical difficulties of the production and acceleration schemes.

Section 4.4 will discuss beta beams based on different ions than $^6$He and $^{18}$Ne: $^8$B and $^8$Li. This configuration has been proposed thanks to an innovative scheme of ion production, see Section 2.3.5, capable in principle og producing much higher radioactive ion fluxes than the standard ISOL techniques used for $^6$He and $^{18}$Ne. $^8$B and $^8$Li having higher $Q_\beta$ values than $^6$He and $^{18}$Ne can produce higher energy neutrino beams for the same accelerator setup, at the price of a smaller collimation (that depends on $\gamma$). This latter development suggested several studies on high energy $^8$B /$^8$Li setups, capable of producing neutrino beams with enough energy to cover the so-called "magic baseline", the optimal configuration to study neutrino mass hierarchy. These latter configurations will be discussed in Section 4.5.

## 4.2   High Energy Beta Beams

High energy beta beams (HEBB) have been introduced by [125], where setups different from the original beta beam concept have been proposed for the first time.

This setup is the same as CFBB, except for the fact that the final accelerator is designed to accelerate $^6$He up to $\gamma = 350$ (2.3 times higher that the maximum $\gamma(^6$He) reachable at the SPS), a condition fulfilled by an accelerator capable of accelerating protons at 1 TeV.

The same number of ion decays/year as the CFBB has been considered.

Two major upgrades of the accelerator scheme are needed for high energy beta beams. Of course a new accelerator is needed. Proton accelerators at 1 TeV energy have been recently dismantled (HERA at Desy) or are going to be shut-down (Tevatron at Fermilab).

The LHC is a collider with a very slow acceleration cycle which makes it unsuitable for the acceleration of the large number of radioactive ions required for a beta beam. A possible energy upgrade of the LHC would require a new higher energy injector, SPS+ [71], which could be used for a higher energy beta-beam.

Also the decay ring is heavily affected by a $\gamma$ increase of the stored ions. First its length scales linearly with $\gamma$, since the magnetic rigidity of the ions is proportional to $\gamma$ and the fraction of length of the straight decay section cannot be reduced without compromising the neutrino fluxes at the far detector. Second the number of ions stored in the decay ring scales again with $\gamma$, according to the Lorentz boost on their lifetime. This discussion has already been developed in Section 2.

Coming back to physics performances, another important advantage of a high energy beta beam is the possibility to increase the baseline length to the point where sensitivity to sign($\Delta m^2_{23}$) becomes sizable. Matter effects are a double edged sword in this kind of studies: in the region where sensitivity to sign($\Delta m^2_{23}$) is great enough to decouple CP effects and matter effects (as discussed in Section 1.2 matter effects produce CP fake signals that can act as a background for genuine CP effects) the experiment has access to the two important signals; on the contrary, in the parameter region where they cannot be decoupled, the sign($\Delta m^2_{23}$) degeneracy tends to reduce CP sensitivity.

The claim that HEBB do have a better energy reconstruction than CFBB and for this reason they should have a better rejection of the neutrino oscillation parameter degenerate solutions seems less robust. While it is true that energy resolution is almost constant at low energy because it is dominated by Fermi motion, producing a smaller number of useful energy bins in CFBB than in HEBB. Nevertheless the fraction of non-quasi-elastic events in HEBB is higher, and these events are reconstructed with wrong energy in a water Čerenkov detector, weakening the benefit of a higher number of energy bins. The greater power of HEBB in rejecting degenerate solutions is derived in turn from the higher statistics.

Reference [144] studies the case of a water Čerenkov detector at $\gamma = 350$ for $^6$He and $^{18}$Ne ions. In a water Čerenkov detector only quasi-elastic (QE) events can be properly reconstructed, so by increasing the average neutrino energy, the fraction of well reconstructed events decreases, until the point where the flux increase provided by the higher gamma is vanished by the loss of QE events. According to [144] (see Fig. 4.1) this happens for $\gamma \simeq 400$.

Backgrounds from NC are much more in HEBB than in CFBB, but they cluster at small energies. As demonstrated by Ref. [144], a simple lower cut in the visible energy keeps NC backgrounds to a tolerable level. Also atmospheric neutrinos integrated in the signal energy range increase, but much less than signal events, when compared to CFBB. This feature implies that in HEBB the bounds to the beta beam duty cycle derived from the atmospheric neutrino background rate are less severe, allowing for higher duty cycles.

Following the results of [144] (see Fig. 4.2) a $\gamma = 350$ beta beam would have marginal improvement as far as $\theta_{13}$ and LCPV sensitivities are concerned with respect to an SPS-based beta beam at the maximum $\gamma$ ($\gamma = 150$

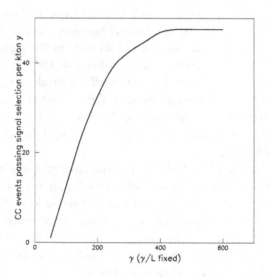

Fig. 4.1   Number of useful events in a 500 kt water Čerenkov detector placed at the first oscillation maximum of a high energy beta beam as a function of the parent ion $\gamma$. From [144].

for $^6$He and $^{18}$Ne ) and at the optimal baseline (L=300 km) [1], while $\gamma = 350$ has definitely better performances as far as sign($\Delta m^2_{23}$) sensitivity is concerned. In this aspect anyway no study has been done so far about sensitivity combining beta beam with atmospherics in this configuration.

It should also be noted that the constant binning chosen in this study for the different setups (with the first bin ranging from 0 to 500 MeV) penalizes very much the lower $\gamma$ configurations, wich are reduced to a counting experiment.

Following the above discussion, a water Čerenkov detector shows some limitation in the energy range of high energy beta beam, if only quasi-elastic events can be efficiently reconstructed. To overcome this problem different detector technologies have been taken into account for HEBB.

In [145], the case of a totally active scintillating detector (TASD), derived from the NO$\nu$A project, has been considered.

---

[1] This is the right way to compare the two options, by fixing the optimal baseline in the two cases. The problem is then the practical possibility of actually having a megaton class detector at the right depth at the optimal baseline.

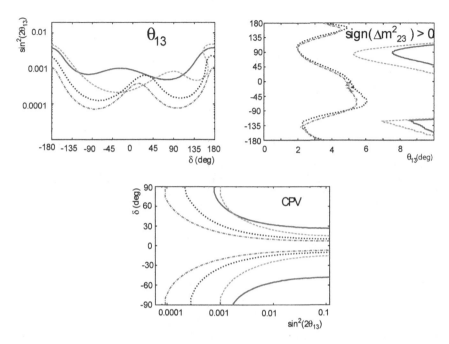

Fig. 4.2   Upper-left panel: exclusion plot for $\theta_{13}$ at 99% CL for the $\gamma_{6\,\mathrm{He}} = 60$, $\gamma_{18\,\mathrm{Ne}} = 100$ and L=130 km setup (solid), the $\gamma = 120$ and L=130 km setup (dashed), the $\gamma = 150$ and L=300 km setup (dotted) and the $\gamma = 350$ and L=730 km (dashed-dotted). Upper-right panel: region on the plane ($\theta_{13}$, $\delta_{CP}$) in which $\mathrm{sign}(\Delta m^2_{23})$ can be measured at 99% CL for the true $\mathrm{sign}(\Delta m^2_{23}) = +1$ and the same four setups. Lower-center: CP-violation exclusion plot at 99% CL for the same four setups. The solar and atmospheric parameters are fixed to their present best fit values and the discrete ambiguities are assumed to be resolved. From [144].

It should be noted that the needs of a beta beam experiment, a tiny $\nu_\mu(\overline{\nu}_\mu)$ signal in an intense $\nu_e(\overline{\nu}_e)$ beam, are different from super beam experiments like NOνA, where a tiny $\nu_e(\overline{\nu}_e)$ signal has to be identified in an intense $\nu_\mu(\overline{\nu}_\mu)$ beam. In particular the particle identification of charged pions against muons is very different from the identification of neutral pions against electrons. So background rates and signal efficiencies cannot be directly derived from the studies on super beam experiments and should be recomputed specifically for beta beam experiments. Overall performances depend very much upon background rates, especially in the small $\theta_{13}$ regime, and cannot be reliably computed in the absence of a full simulation (and possibly a full reconstruction of the events).

Reference [145] lacks such a simulation, and compares water Čerenkov detectors of 500 kt mass at $\gamma = 200$ (Setup I) with TASD detectors of 50 kt at $\gamma = 500$ (Setup II) and $\gamma = 1000$ (Setup III). Regarding $\theta_{13}$ and LCPV sensitivity, Setups I and II have similar performances, assuming constant decay rates in every setup, while Setups II is clearly better as far as sign($\Delta m_{23}^2$) sensitivity is concerned (but again the combination with atmospherics, not possible for a TASD surface detector, is not taken into account). We will not dwell too much on performances of Setup III which requires LHC as an accelerator.

Also shown in [145] is the scaling of performances with the number of ion decays/year (Fig. 4.3). Assuming a scaling law as:

$$N^i = N_0 \cdot \left(\frac{\gamma_0}{\gamma}\right)^n \qquad (4.1)$$

where $N$ ($N_0$) is the number of decays/year at a given $\gamma$ ($\gamma_0$), and $\gamma_0$ is a reference point. $n = 0$ is the case of constant ion decays/year, while $n = 1$ is the case of "constant power" in the accelerator. In this latter case the sensitivity of the setup becomes rather independent from $\gamma$, showing that the assumptions about this scaling law are very important for the overall comparisons.

A different detector technology has been considered in [146]: an iron calorimeter, where the sensitive elements (2 cm thick glass RPC planes with a 2 mm gas filled gap) are interleaved with iron plates (4 cm thick). This configuration has the advantage of providing a higher density than a TASD detector, such that a 40 kt detector could fit a present LNGS hall, a very attractive experimental situation. A full simulation of this detector has been performed, allowing for a robust sensitivity estimation. The fraction of NC backgrounds with respect to the non-oscillated $\nu_e$ events is $8.8 \times 10^{-3}$ at $\gamma = 580$, a much higher rate than the $10^{-3}$ rate assumed at $\gamma = 500$ for a TASD detector. It is expected that a totally active detector can have a better background rejection than an iron calorimeter, nevertheless the discrepancy of almost one order of magnitude seems to indicate some underestimation of the background rates in [145].

Overall performances of this setup almost match those of the CERN-Fréjus scenario, again assuming a constant ion decay rate. Combined sensitivity with atmospheric neutrinos of this setup have been also studied in [146] showing that its sensitivity, as expected from the longer baseline, outperforms CFBB performances.

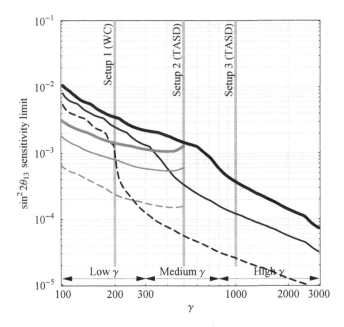

Fig. 4.3 The final $\sin^2 2\theta_{13}$ sensitivity limit (including systematics, correlations, and degeneracies) as function of $\gamma$ for $L = 1.3\,\gamma$ (TASD) and $L = 2.6\,\gamma$ (WC). The bands correspond to the Totally Active Scintillator Detector (TASD, green/light shaded region) and the water Čerenkov detector (WC, blue/dark shaded region) at the $3\sigma$ confidence level, where the isotope luminosity exponent $n$ is varied from 0 (lower ends of the bands) over 0.5 (black curves) to 1 (upper ends of the bands). The vertical lines correspond to our low-, medium-, and high-$\gamma$ setups. From [145].

## 4.3 Monochromatic Neutrino Beams

Monochromatic neutrino beams based on the electron capture process (ECB) have been discussed in Section 2.7.5; here possible performances under the assumption of parent ion decay rates of $10^{18}$ decays/year will be discussed.

ECB are certainly an intriguing experimental setup, but for LCPV searches they have two major apparent limitations: there is no way to have antineutrino beams (a conceptual possibility for the production of monochromatic neutrino beams is discussed later in this section) and they miss spectral information, which is very important to solve degeneracies.

To overcome these limitations interesting experimental strategies have been introduced.

In [118, 147] it has been proposed to study $\theta_{13}$ and LCPV in a ECB setup based on the $^{150}$Dy ion (3.1 min lifetime and $Q = 1794$ keV) running the beam at two $\gamma$s tuned to the first and the second oscillation maximum. Two setups are considered, the first, based on the SPS and the CERN-Fréjus baseline, would run the ions at $\gamma = 90$ and $\gamma = 195$ the second, based on SPS+, considers $\gamma = 195, 440$. The detector is a water Čerenkov of 440 kt in both cases. Performances of ECB in these configurations are very promising. It should anyway be noted that in this study any background contamination has been neglected and 100% signal efficiency has been assumed, a quite optimistic scenario.

Reference [148] proposes a more aggressive strategy. In a beta beam the neutrino energy has a dependence on the radial position $R$ of the interaction point:

$$E_\nu(R) \simeq \frac{2\gamma Q_\beta}{1 + (\gamma R/L)^2} . \tag{4.2}$$

As shown in Fig. 4.4, this relation has very little impact on the CFBB setup, but could have a sizable effect on the setup chosen by the authors of reference [148]: $^{110}_{50}Sn$ isotopes with $Q = 267$ Kev and a 4.11 h lifetime. Running these ions at $\gamma = 2500$ and a baseline of 600 km, the energy of events at $R = 100$ m would have a 15% smaller energy than events at $R = 0$. The vertex resolution of a water Čerenkov detector, about 30 cm,

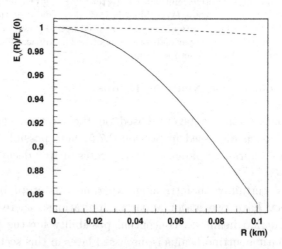

Fig. 4.4 Fractional variation of the neutrino energy $E_\nu$ as function of the radial distance from the beam axis. Dotted line: $^6$He with $\gamma = 100$ and a baseline of 130 km; solid line: $^{110}_{50}$Sn with $\gamma = 2500$ and a baseline of 600 km.

Table 4.1   Characteristics of $^8$B and $^8$Li compared with $^6$He and $^{18}$Ne.

| $\beta^+$ emitters | | | $\beta^-$ emitters | | |
|---|---|---|---|---|---|
| Ion | $Q_{\text{eff}}$ (MeV) | Z/A | Ion | $Q_{\text{eff}}$ (MeV) | Z/A |
| $^{18}$Ne | 3.30 | 5/9 | $^6$He | 3.51 | 1/3 |
| $^8$B | 13.92 | 5/8 | $^8$Li | 12.96 | 3/8 |

is suitable for such a measurement. The requirements of beam divergence $p_x/p_z \lesssim 1$ $\mu$rad, and an equivalent precision of the absolute beam direction seem anyway very challenging.

Under these extreme conditions this configuration could allow very good energy resolution and a finite beam energy range. According to [148] it could reach excellent $\theta_{13}$ and LCPV sensitivities.

A way to generate monochromatic antineutrino beams has been delineated in [121] (see also Section 2.7.5). It is based on the process of the bound-state $\beta$ decay [149] where the electron is created in a previously unoccupied bound atomic state and the antineutrino is emitted at a fixed energy.

Candidates exist like $^{108}_{47}Ag^{46+}$ with $\tau_{1/2} = 24.4$ s and neutrino energies of 1.90 and 1.67 MeV for the EC and bound-beta lines respectively, but it should be noted that the branching ratios for such processes are of about 1%, making it very difficult even conceptually to produce significative neutrino fluxes.

## 4.4   Beta Beams Based on $^8$B and $^8$Li Ions

$^8$B and $^8$Li ions have a significantly higher $Q_\beta$ value than $^6$He and $^{18}$Ne as can be derived from Table 4.1. In Section 3.1 it has already been shown that higher $Q_\beta$ ions can allow greater neutrino energies for the same $\gamma$:

$$E_\nu^{\text{max}} = 2\gamma Q_\beta. \tag{4.3}$$

Furthermore the Z/A of the $^8$B /$^8$Li ions are higher than the formers': such that considering the $\beta^-$ emitters they could produce a neutrino beam 4.74 times more energetic than a $^6$He /$^{18}$Ne beam, for the same accelerator energy, with a shorter decay ring length. On the other hand the merit factor of a $^8$B /$^8$Li beam (see Section 3.1) is smaller than a $^6$He /$^{18}$Ne beam since it is inversely proportional to $Q_\beta$ and so it would produce smaller fluxes at the same neutrino energy.

In [93], as discussed in Section 2.3.5, an innovative procedure has been proposed to produce $^8$B$/^8$Li ions, in principle capable of producing 2 - 3 orders of magnitude more radioactive ion fluxes. An increase of 2 - 3 orders of magnitude of the ion fluxes to be accelerated requires important R&D on the accelerators to cope with the larger number of charges for acceleration, accumulation and storage.

The physics case of a $^8$B$/^8$Li beta beam based on the Fermilab Main Injector has been proposed in [150]. The Main Injector is capable of accelerating $^8$B up to $\gamma = 80$ and $^8$Li up to $\gamma = 48$ such that a 50 - 100 kt liquid argon detector could be put in place at Soudan (732 km baseline). The liquid argon TPC technology is very appealing for neutrino beams with energy greater than 1 GeV, since it can reconstruct with great precision neutrino events of any multiplicity. In the case of beta beams charged pion identification against muons is a key ingredient. We have seen that in water the most powerful tool to positively identify muons is the detection of their decay chain. In argon this method cannot be used because given the higher $Z$ of the nucleus, the probability of muon absorption before decay is very high, about 73%, in contrast with the probability of absorption in water, only 22%.

In liquid argon NC rejection can be performed thanks to the measurement of both the momentum and the range of the track and the identification of the nuclear fragments generated in the nuclear capture of the pion. It has been suggested in [150] that in this way a $10^4$ rejection of charged pions can be achieved.

The authors of [151] have studied the case of a mixed $^8$B$/^8$Li and $^6$He$/^{18}$Ne beta beam, based on SPS. A 500 kt water Čerenkov detector with a baseline of about 700 km would receive the $^8$B$/^8$Li beta beam at the first oscillation maximum and the $^6$He$/^{18}$Ne beta beam at the second oscillation maximum. The same ion decays/year of CFBB are assumed also for $^8$B and $^8$Li.

This combination of first/second maximum is very powerful in solving degeneracies, since the differences between the oscillation patterns of the different oscillation components are more and more visible with the development of oscillations. Nevertheless at the second oscillation maximum fluxes are reduced by about one order of magnitude, and statistics is the major component in sensitivity to $\theta_{13}$ and LCPV. Therefore this has little advantage as far as $\theta_{13}$ and LCPV are concerned, while it outperforms CFBB as far as sign($\Delta m_{23}^2$) sensitivity is concerned.

Along this line it is also interesting to note the study of reference [152] where the case of a single $^{18}$Ne exposure is considered at $\gamma = 450$ (within the reach of the SPS+) and with a 50 kt iron detector placed at a baseline of 1050 km (CERN-Boulby mine). This neutrino-only setup would cover both the first and the second oscillation maximum. While the $\theta_{13}$, LCPV and sign($\Delta m_{23}^2$) sensitivities do not ouperform those of other beta beam setups, this particular scheme could reach an interesting sensitivity to the octant of $\theta_{23}$.

## 4.5   High Energy $^8$B /$^8$Li Beta Beams

The combination of high energy, $^8$B /$^8$Li based, beta beams allows the so called "magic baseline" $L_{\text{magic}}$ to be covered.

The concept of a magic baseline [153, 35] derives from the observation that in Eq.(1.5) for $\rho L = \sqrt{2}\pi/G_F Y_e$ ($Y_e$ is the electron fraction inside the earth) any $\delta_{\text{CP}}$ dependence disappears from $P_{e\mu}$ allowing sign($\Delta m_{23}^2$) effects to be measured without any degenerate solution.

The measurement of neutrino oscillation at the magic baseline is the ideal complement to LCPV searches, since it decouples fake CP effects generated by matter effects from the genuine CP effects looked for in LCPV searches.

According to the Preliminary Reference Earth Model PREM [154] earth matter density profile, $L_{\text{magic}} \simeq 7690$ km, the resonance energy for matter effects would be:

$$E_{\text{res}} \equiv \frac{|\Delta m_{31}^2| \cos 2\theta_{13}}{2\sqrt{2}G_F N_e} \simeq 7 \text{ GeV} \tag{4.4}$$

for $|\Delta m_{31}^2| = 2.4 \cdot 10^{-3}$ eV$^2$ and $\sin^2 2\theta_{13} = 0.1$.

It is important to note that close to matter resonance, the flux of oscillated events at the detector roughly falls as a function of $1/L$ (against the $1/L^2$ fall of vacuum oscillations), which means that longer baselines might be preferred.

Studies of beta beams at the magic baselines have been initiated [155] within the context of the India-based Neutrino Observatory (INO) [156], where a large magnetized iron calorimeter (ICAL) is set to come up. The detector is designed to be 50 kt, made of 6 cm thick iron slabs interleaved with glass RPCs. The CERN-INO distance approaches the magic baseline, being 7152 km.

The authors of [155] consider a $^8$B$/^8$Li beta beam with $\gamma_{^8\text{Li}} = 250\text{-}500$ (max $\gamma_{^8\text{Li}}$ attainable at the SPS is 168.75). The production of neutrinos with energy $E_{\text{res}} = 7$ GeV requires $^8$Li ions to be accelerated at $\gamma_{^8\text{Li}} \geq 225$. At the SPS+ energies, $\gamma_{^8\text{Li}} = 350$, this project could measure $\sin^2 2\theta_{13} \simeq 0.01$ (90% CL) and measure $\text{sign}(\Delta m_{23}^2)$ at $3\sigma$ if $\theta_{13} \geq 1.2 \cdot 10^{-2}$.

Background rejection is essential to guarantee such performances. In [155] NC backgrounds are not taken into account, while in a similar setup, as discussed Section 4.2, NC backgrounds are quite severe under a full detector simulation.

The magic baseline is the ideal complement to searches focused on LCPV. The authors of [157] studied a beta beam configuration where a triangular shaped decay ring would send neutrinos generated by $^8$B and $^8$Li ions accelerated at $\gamma = 350$ to two 50 kt iron calorimeters (with the same characteristics as ICAL-INO) at two baselines of 2000 and 7000 km. In this configuration one detector is optimized to LCPV and $\theta_{13}$ searches and the other to $\text{sign}(\Delta m_{23}^2)$. The combination of the two is synergic, more powerful than any configuration with the two detectors at the same baseline. Again the background rejection of such detectors is quite optimistic, the NC fractional background being estimated at $10^{-5}$. Performances are studied for a range of possible ion decay rates. For a $^8$B$/^8$Li decay rate five times higher than the $^6$He$/^{18}$Ne decay rate this setup would match the high energy beta beam option of [157] as far as $\theta_{13}$ and LCPV sensitivities are concerned, while $\text{sign}(\Delta m_{23}^2)$ sensitivity would be better in the $^8$B$/^8$Li configuration.

In [158] a comprehensive comparison of $^8$B$/^8$Li and $^6$He$/^{18}$Ne beta beams have been performed. The authors consider several terms of comparison, we will summarize here the comparison at constant $\gamma$ (this comparison is not exactly the same as "constant accelerator complex" because, as discussed in Section 4.4, the different $Z/A$ ratios of the different couples of ions bring different possible values of $\gamma$, keeping the accelerator constant). The comparison is performed keeping the detector the same, a 50 kt INO-like iron magnetized detector, which is basically devoted to studying performances at small values of $\theta_{13}$. A first conclusion is that the simple scaling laws that can be obtained by the $Q_\beta$ values of the ions about the useful decays $N_\beta$ and $\gamma$ to keep equal the beam performances (see Section 4.4):

$$N_\beta^{B+Li} \simeq 12 \cdot N_\beta^{Ne+He} \qquad \gamma^{Ne+He} \simeq 3.5 \cdot \gamma^{B+Li}$$

is confirmed by the detailed computations.

An optimal choice for a single setup is not possible, "short base-line" (700 - 1000 km) $^6$He$/^{18}$Ne beta beams outperform "magic baseline" $^8$B$/^8$Li beta beams as far as LCPV sensitivity is concerned (as expected since the magic baseline is not, by construction, sensitive to LCPV), while the contrary is true for sign($\Delta m^2_{23}$) sensitivity. The two configurations have similar $\theta_{13}$ sensitivities.

It should be noted that detector performances might strongly bias these conclusions. In [158] the performances of the INO detector are parameterized in the absence of a full simulation and kept constant in the whole energy range studied in the paper. In particular NC rejection cannot be constant at different neutrino energies, and the $10^{-4}$ NC rejection factor considered in the paper is in disagreement with the rejection factor computed with a full simulation in [146] for one of the $^6$He$/^{18}$Ne $\gamma$ values considered in the comparison. The energy threshold of 1 GeV for the muon detection of the INO-like detector is not well matched to the neutrino energies in setups with $^6$He$/^{18}$Ne at $\gamma = 350\text{-}500$, and so the "short baselines" are not optimally exploited with such a detector. Indeed, following [158], a 500 kt water Čerenkov detector at the "short baselines" would have far better LCPV and $\theta_{13}$ performances with respect to the 50 kt iron magnetized detector.

It is evident that the two setups studied in [158] complement each other and that the optimal beta beam option should include both of them. This configuration will be discussed in the following section.

## 4.6   Comparison with Other Neutrino Facilities and Green-field Scenarios

We have seen in the previous sections that the comparison of different beta beam setups is severely influenced by assumptions about ion decay rates, detector performances and systematic errors.

The situation gets even worse in trying to compare beta beam setups with other setups derived from different neutrino beam concepts. In this case it is very difficult to have a homogeneous treatment of the many assumptions necessary to produce a sensitivity estimation.

The ISS study group [28] made a very succesful effort to produce a fair comparison between the different facilities. The performances of the different setups were compared by using the same analysis tools (Globes [135]) and reaching a consensus about the present knowledge of the performances of the different setups. The considered setups are the T2HK, SPL and WBB super beams, the beta beam configurations of CFBB and of $\gamma = 350$

in Section 4.2 and conservative and optimized neutrino factory setups. Not considered in the comparison are the performances of $^8$B $^8$Li beta beam setups discussed in Section 4.5, published after the ISS study.

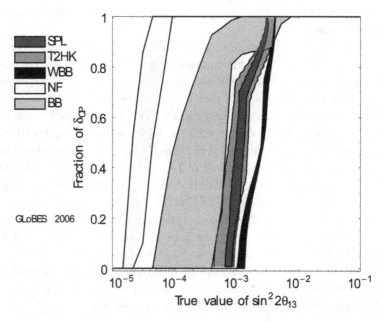

Fig. 4.5   The discovery reach at 3 $\sigma$ level for different facilities in $\sin^2 2\theta_{13}$. The discovery limits are shown as a function of the fraction of all possible values of the true value of the CP phase $\delta$ ('Fraction of $\delta_{CP}$') and the true value of $\sin^2 2\theta_{13}$. The right-hand edges of the bands correspond to the conservative setups while the left-hand edges correspond to the optimized setups. From [28].

Fig. 4.5, shows the discovery reach of the various facilities in $\sin^2 2\theta_{13}$. The figure shows the fraction of all possible values of the true value of the CP phase $\delta$ ('fraction of $\delta_{CP}$') for which $\sin^2 2\theta_{13} = 0$ can be excluded at the $3\sigma$ confidence level as a function of the true value of $\sin^2 2\theta_{13}$. Of the super beam facilities, the most sensitive is the T2HK with the optimized parameter set. The SPL super beam performance is similar to that of T2HK, while the performance of the WBB is slightly worse. The limit of sensitivity of the super beam experiments is $\sim 5 \times 10^{-4}$; for $\sin^2 2\theta_{13} \gtrsim 10^{-3}$ the super beam experiments can exclude $\sin^2 2\theta_{13} = 0$ at the $3\sigma$ confidence level for all values of $\delta$. The CFBB beta beam setup has good sensitivity to $\sin^2 2\theta_{13}$ for $\sin^2 2\theta_{13} \sim 10^{-3}$, but decreases in sensitivity for values of

$\theta_{13}$ to just under the sensitivity limit of T2HK. The $\gamma = 350$ beta beam has significantly better performance, with a sensitivity limit of $\sin^2 2\theta_{13} \gtrsim 5 \times 10^{-5}$. Both the conservative and the optimized neutrino factory setups have a significantly greater $\sin^2 2\theta_{13}$ discovery reach; the optimized setup having a sensitivity limit of $\sim 1.5 \times 10^{-5}$.

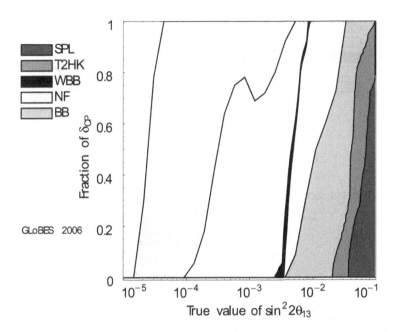

Fig. 4.6   The discovery reach at 3 $\sigma$ level for different facilities in $\mathrm{sign}(\Delta m_{23}^2)$. The discovery limits are shown as a function of the fraction of all possible values of the true value of the CP phase $\delta$ ('Fraction of $\delta_{CP}$') and the true value of $\sin^2 2\theta_{13}$. The right-hand edges of the bands correspond to the conservative setups while the left-hand edges correspond to the optimized setups. From [28].

Fig. 4.6 shows the discovery reach of the various facilities in $\mathrm{sign}(\Delta m_{23}^2)$. The various bands shown in the figure have the same meaning as those shown in Fig. 4.5; the discovery reach is again evaluated at the $3\sigma$ confidence level. Of the super beam setups considered only the WBB has significant sensitivity to the mass hierarchy with a sensitivity limit of $\sin^2 2\theta_{13} \gtrsim 3 \times 10^{-3}$. Of the beta beam setup only the $\gamma = 350$ option with the relatively long baseline of 730 km is competitive with the WBB , having a comparable sensitivity limit. It is worth noting that the combination with atmospheric neutrinos is not taken into consideration in those plots. The neutrino factory, benefiting from the long baseline, outperforms the other facilities, the sensitivity limit of the conservative option being $\sin^2 2\theta_{13} \gtrsim 1.5 \times 10^{-4}$, while the sensitivity limit of the optimized facility is $\sin^2 2\theta_{13} \gtrsim 1.5 \times 10^{-5}$.

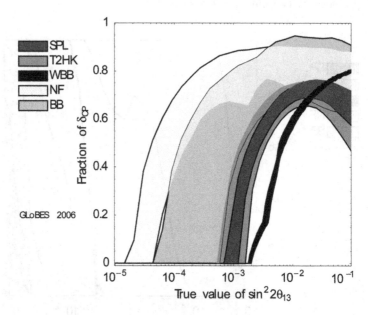

Fig. 4.7  The discovery reach at 3 $\sigma$ level for different facilities in leptonic CP violation sensitivity. The discovery limits are shown as a function of the fraction of all possible values of the true value of the CP phase $\delta$ ('Fraction of $\delta_{CP}$') and the true value of $\sin^2 2\theta_{13}$. The right-hand edges of the bands correspond to the conservative setups while the left-hand edges correspond to the optimized setups. From [28].

Fig. 4.7 shows the discovery reach of the various facilities in the CP phase $\delta$. The T2HK and the SPL super beams show a greater sensitivity to CP violation for $\sin^2 2\theta_{13} \sim 10^{-3}$ than the WBB experiment. However, the WBB experiment has sensitivity for a larger range of values of $\delta$ that the other super beam facilities considered for $\sin^2 2\theta_{13} \sim 10^{-1}$. The performance of the CFBB beta beam is comparable to that of the optimized T2HK experiment. The high energy ($\gamma = 350$) beta beam shows considerably better performance: a sensitivity limit of $\sim 4 \times 10^{-5}$ and a CP coverage of around 90% for $\sin^2 2\theta_{13} \gtrsim 10^{-2}$. For low values of $\theta_{13}$, $\sin^2 2\theta_{13} \lesssim 10^{-4}$, the conservative neutrino factory performance is comparable with that of the high energy beta beam. For larger values of $\theta_{13}$, the CP coverage of the high energy beta beam is significantly better. The optimized neutrino factory outperforms the optimized beta beam for $\sin^2 2\theta_{13} \lesssim 4 \times 10^{-3}$. For larger values of $\theta_{13}$ the optimized beta beam has a slightly larger CP coverage.

As it is evident from the above figures, different values of $\theta_{13}$ bring different choices about the preferred setup. While it is evident that for very small values of $\theta_{13}$ a neutrino factory setup is preferable (but a full simulation of the detectors is needed in order to convincingly demonstrate the extreme background rejection needed), with $\theta_{13}$ in the reach of next generation experiments (see Section 1.3.3), the situation becomes much more complicated.

As a final, important remark, there are decisive factors that should be taken into consideration when comparing different facilities and that were not available to the ISS studies. These are for instance the final cost of the facility, the timescales to completion of the needed R&D studies and the construction of the facility and the detectors, a fair comparison of the assumptions about the machines in the different setups.

After the ISS study the neutrino factory setup has been further developed in the International Design Study (IDS) [84]. On the beta beam side, several independent papers explored the ultimate sensitivities of beta beam setups relaxing the constraint of ion decay rates per year. While the decay rates assumed for the CFBB setup are considered already challenging, it is true that they are fulfilled for an incident proton beam power of less than 0.2 MW, while Neutrino and super beams are normalized to an incident proton beam power of 4 MW.

Here we will summarize the findings of two notable papers exploring very different assumptions about the true value of $\theta_{13}$.

reference [160] focused on beta beam performances at the smallest $\theta_{13}$, considering a two-detector 50 kt iron magnetized INO-like setups, at two baselines of 730 and 7150 km.

The main question of the paper could be summarized as "Which performances are required for a beta beam setup in order to match the sensitivities of the IDS neutrino factory?"

The answer is that a setup alternating a $\gamma = 650$ $^8$B/$^8$Li run with a $\gamma = 575$ $^6$He/$^{18}$Ne run, a total luminosity of $10\times$ the nominal beta beam luminosity could match the sensitivities of the IDS neutrino factory as far as $\theta_{13}$, sign($\Delta m_{23}^2$)and LCPV sensitivity are concerned (see Table 4.2).

The study published in [161] compared possible beta beam setups with super beams and neutrino factories in the specific scenario where the true value of $\theta_{13}$ is in the reach of Double Chooz ($\sin^2 2\theta_{13} \geq 0.04$ ($3\sigma$)). The outcome of the study is summarized in Table 4.3, which displays the facilities capable of measuring i) $\sin^2 2\theta_{13} > 0$ at $5\sigma$ ii) mass hierarchy at $3\sigma$ for any value of $\delta_{CP}$ and iii) LCPV at $3\sigma$ for 80% of the allowed values of $\delta_{CP}$. Being a study oriented to beta beams, the table shows the lowest values of $\gamma$ for a beta beam to reach these performances at four possible baselines (corresponding to CERN-LNGS or FNAL-Soudan: 730 km, FNAL-Ash River: 810 km, CERN-Boulby or J Parc-Korea: 1050 km and FNAL-Homestake: 1290 km). The mass hierarchy requirement rules out the CFBB setup even considering the atmospherics. The luminosity scaling factor $\mathcal{L}$ introduced here is the product of useful ion decays per year × running time × detector mass × detection efficiency, $\mathcal{L}=1$ corresponds to $2.9(1.1) \cdot 10^{18}$ decays/year× 10 years × 500 (100) kt for water (TASD) detectors × default signal efficiency.

From the table it is evident that a beta beam setup with a modest gain in luminosity with respect to the nominal one, could fulfill all the three benchmarks and outperform any other setup. Also an SPS based setup with $\gamma = 150$, $L = 400, 600$ km and $\mathcal{L} \geq 2$ could fulfill these benchmarks.

## 4.7 Conclusions

The summary of this chapter is probably that more questions have been asked than answers given. The field of beta beam setups has grown very fast in the few past years, developing several new concepts for very aggressive experimental setups.

Among the main questions are which kind of new accelerators (if any) will be developed at CERN for the LHC upgrade. This choice will drive the development of high energy beta beams. Also the ultimate intensities of ion decays in a beta beam setup could bring very important improvements in beta beam performances, and the studies about the production of the $^8$B and $^8$Li ions have just begun.

The challenging possibility of monochromatic neutrino beams could bring a very powerful tool to the ultimate quest of measuring the unknown neutrino oscillation parameters.

While the performances of water Čerenkov detectors are quite well known, different detector technologies that are needed in case of high energy beta beams are less known, such as iron magnetized detectors, totally active scintillator detectors and liquid argon detectors. Detailed studies about performances of such detectors in the very specific beta beam setups could clarify the path of optimization for the different facilities.

What can be said is that beta beams setups have all the potential to be the ultimate facility of neutrino physics and that in the very welcome situation where $\theta_{13}$ is within the reach of next generation experiments, beta beams are the best candidate to complete the measurement of the still unknown today $\theta_{13}$, sign$(\Delta m_{23}^2)$ and $\delta_{\rm CP}$ parameters.

Table 4.2 Performances of various beta beam setups at $3\sigma$ in addressing the key unsolved issues: mass ordering, CP violation and $\sin^2 2\theta_{13}$ sensitivity reach. For CP sensitivity and mass ordering, the minimum values of $\sin^2 2\theta_{13}$ (true) required for a positive conclusion are presented. Results are shown for a five-year run with the reference luminosity: $1.1 \times 10^{18}$ $(2.9 \times 10^{18})$ useful ion decays per year in the $\nu$ $(\bar{\nu})$ mode as well as one order of magnitude higher statistics. The numbers without (with) parantheses correspond to $\delta_{CP}$ (true) $= 90°$ $(\delta_{CP}$ (true) $= 270°)$. Note that the $\sin^2 2\theta_{13}$ sensitivity reach is independent of the value of $\delta_{CP}$ (true) and the true mass ordering because the prospective "data" have been generated at $\theta_{13} = 0$. The CERN-INO baseline is insensitive to $\delta_{CP}$. For comparison, the expectations from an optimized two-baseline neutrino factory setup with upgraded magnetized iron detectors are also listed [164].

| Setup | Mass Ordering ($3\sigma$) NH (True) | | CP Sensitivity ($3\sigma$) NH (True) | | $\sin^2 2\theta_{13}$ Sensitivity ($3\sigma$) | |
|---|---|---|---|---|---|---|
| | $1.1\times10^{18}$ & $2.9\times10^{18}$ | $1.1\times10^{19}$ & $2.9\times10^{19}$ | $1.1\times10^{18}$ & $2.9\times10^{18}$ | $1.1\times10^{19}$ & $2.9\times10^{19}$ | $1.1\times10^{18}$ & $2.9\times10^{18}$ | $1.1\times10^{19}$ & $2.9\times10^{19}$ |
| CERN-INO $\gamma = 650$, 7152 km | $4.7\times10^{-4}$ $(4.9\times10^{-4})$ | $9.4\times10^{-5}$ $(1.2\times10^{-4})$ | Not possible | Not possible | $1.14\times10^{-3}$ | $1.76\times10^{-4}$ |
| CERN-LNGS $\gamma = 575$, 730 km | $3.89\times10^{-3}$ $(9.23\times10^{-3})$ | $1.58\times10^{-3}$ $(4.48\times10^{-3})$ | $1.6\times10^{-4}$ $(1.8\times10^{-4})$ | $1.97\times10^{-5}$ $(2.03\times10^{-5})$ | $1.78\times10^{-3}$ | $8.59\times10^{-5}$ |
| CERN-BOULBY $\gamma = 575$, 1050 km | $2.49\times10^{-3}$ $(7.87\times10^{-3})$ | $2.19\times10^{-4}$ $(4.1\times10^{-3})$ | $1.85\times10^{-4}$ $(2.02\times10^{-4})$ | $1.99\times10^{-5}$ $(2.04\times10^{-5})$ | $1.41\times10^{-3}$ | $1.45\times10^{-4}$ |
| CERN-LNGS $\gamma = 575$, 730 km + CERN-INO $\gamma = 650$, 7152 km | $2.7\times10^{-4}$ $(3.58\times10^{-4})$ | $4.64\times10^{-5}$ $(5.45\times10^{-5})$ | $1.42\times10^{-4}$ $(1.49\times10^{-4})$ | $1.78\times10^{-5}$ $(1.88\times10^{-5})$ | $5.46\times10^{-4}$ | $5.26\times10^{-5}$ |

| | | | | | | |
|---|---|---|---|---|---|---|
| CERN-BOULBY $\gamma = 575$, 1050 km + CERN-INO $\gamma = 650$, 7152 km | $2.67\times10^{-4}$ $(3.37\times10^{-4})$ | $4.57\times10^{-5}$ $(5.17\times10^{-5})$ | $1.63\times10^{-4}$ $(1.76\times10^{-4})$ | $1.8\times10^{-5}$ $(1.87\times10^{-5})$ | $6.1\times10^{-4}$ | $6.69\times10^{-5}$ |
| Optimized neutrino factory setup with two improved golden detectors (50 kt each) placed at $L = 4000$ km & 7500 km respectively. $E_\mu = 20$ GeV & total $5\times10^{21}$ decays for $\mu^-$ & $\mu^+$ each. | | | | | | |
| Optimized neutrino factory (100% of $\delta_{CP}$ (true) coverage) | $4.5\times10^{-5}$ | | $1.5\times10^{-5}$ | | $4.5\times10^{-5}$ | |

Table 4.3 Minimal $\gamma$ (rounded up to the next 10) to measure all of the discussed performance indicators (see text) at a specific baseline (in columns) for the given setups and Double Chooz $\sin^2 2\theta_{13}$ best-fit cases, where $\mathcal{L}$ is the luminosity scaling factor. In addition, a number of super beam upgrades and neutrino factory setups are tested for the same criteria and same simulated values. A label "-" refers to no sensitivity in the discussed $\gamma$ ranges. The best options within each setup and $\sin^2 2\theta_{13}$ case are marked boldface. From [161].

| | $\sin^2 2\theta_{13} = 0.04$ | | | | $\sin^2 2\theta_{13} = 0.08$ | | | |
|---|---|---|---|---|---|---|---|---|
| Setup ↓ Baseline [km] → | 730 | 810 | 1050 | 1290 | 730 | 810 | 1050 | 1290 |
| **Beta beams** | | | | | | | | |
| $(^{18}\text{Ne}, {}^{6}\text{He})$ to WC, $\mathcal{L} = 1$ | **220** | 230 | 290 | 350 | **200** | 210 | 240 | 230 |
| $(^{18}\text{Ne}, {}^{6}\text{He})$ to TASD, $\mathcal{L} = 1$ | - | **300** | 370 | 430 | **300** | 310 | 340 | 380 |
| $(^{18}\text{Ne}, {}^{6}\text{He})$ to WC, $\mathcal{L} = 5$ | 190 | 190 | 190 | 230 | 140 | 140 | 140 | 140 |
| $(^{18}\text{Ne}, {}^{6}\text{He})$ to TASD, $\mathcal{L} = 5$ | 200 | 200 | 220 | 230 | 180 | 180 | **170** | 180 |
| $(^{8}\text{B},{}^{8}\text{Li})$ to WC, $\mathcal{L} = 5$ | - | - | 100 | 130 | 80 | 80 | 100 | 110 |
| $(^{8}\text{B},{}^{8}\text{Li})$ to TASD, $\mathcal{L} = 5$ | - | - | 150 | 190 | - | - | 190 | 190 |
| $(^{8}\text{B},{}^{8}\text{Li})$ to WC, $\mathcal{L} = 10$ | 70 | **70** | 90 | 110 | **60** | 70 | 80 | 90 |
| $(^{8}\text{B},{}^{8}\text{Li})$ to TASD, $\mathcal{L} = 10$ | - | 100 | 130 | 140 | **110** | **110** | 120 | 130 |
| **Super beam upgrades** | | | | | | | | |
| T2KK from [162] | | - | | | | ✓ | | |
| NO$\nu$A* from [162] | | - | | | | - | | |
| WBB-120$_S$ from [162] | | - | | | | ✓ | | |
| **neutrino factories** | | | | | | | | |
| IDS-NF 1.0 from [84] | | ✓ | | | | - | | |
| Low-E NF from [163] | | - | | | | ✓ | | |
| **Hybrids** | | | | | | | | |
| NF-SB from [163] | | ✓ | | | | ✓ | | |

# Chapter 5

# Low Energy Beta Beams

## 5.1 Introduction

Beta beams are the ideal tool for measuring neutrino cross sections, as already discussed in Section 3.3.3. This particular feature has been extensively discussed in the literature for neutrino energies around 100 MeV, where a wide set of interesting non-oscillation neutrino experiments is possible.

It should be noted that at those energies neutrino oscillation experiments seem to be rather problematic for several reasons: by lowering the neutrino energy, neutrino fluxes become smaller and so oscillation sensitivities become less attractive; for neutrino energies of 100 MeV or below appearance experiments cannot be conducted with water Čerenkov detectors because charged-current $\nu_\mu$ interactions are below the muon production threshold while electron neutrino disappearance experiments are not competitive with reactor experiments.

However, there is interest in other areas of physics for pure low energy electron (anti)neutrino beams. In [165] it was proposed to build a low energy facility in the 100 MeV energy range for nuclear structure studies and neutrino-nucleus interactions [165–169], electroweak tests of the Standard Model [170, 165, 171, 172] as well as core-collapse Supernova physics [165, 173, 175].

In this energy range the decay ring characteristics and the detector locations have to be re-optimized, as discussed in Section 5.2. The physics potential of cross section measurements of such facilities is discussed in Section 5.3, while the potential for fundamental interaction studies is presented in Section 5.4.

Most of the material in this chapter is (with the kind permission of the author) based on a topical review published by Cristina Volpe [174] material taken from other publications is explicitly referenced in the following.

## 5.2   Low Energy Setups

To produce a beta beam in the 100 MeV energy range with e.g. $^{6}$He and $^{18}$Ne ions, a boost of $\gamma = 7\text{-}14$ is necessary [165].

The low energy component of a beta-beam facility of any $\gamma$ can in principle satisfy the requirements for a low energy beta beam as discussed in [166] (see also Fig. 5.1). However one has to get rid of the high energy events, which requires an off-axis configuration [176] (Section 5.2.1).

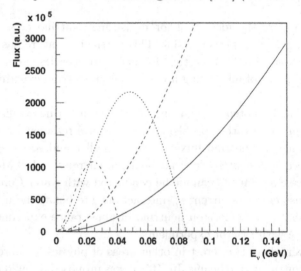

Fig. 5.1   Neutrino fluxes from $^{6}$He beta beams at $\gamma = 7$ (dash-dotted), 14 (dotted), 100 (dashed), 350 (solid) computed on-axis.

A study of the energy spectrum versus intensity issues for various combinations of beta-beam facilities [166] shows that a small devoted storage ring is more appropriate for low energy applications. First conceptual studies of such a storage ring (concerning e.g. size, ion intensities, stacking method, space charge effects) [177, 178] have shown it shares many concerns as those discussed in Chapter 2. In short, it is feasible to construct such a ring and it could possibly achieve higher intensities compared to the higher energy beta beams due to the shorter acceleration time.

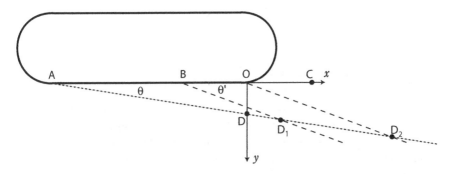

Fig. 5.2  Possible on-axis and off-axis detector locations. See the text for the different ring configurations and the different symbols.

To understand the neutrino fluxes near the decay ring straight sections it is necessary to rewrite Eq. (3.1) making explicit the angular dependence of the neutrino flux $\Phi(E_\nu)$.

The neutrino flux $\Phi_{cm}(E_\nu)$ in the rest (cm) frame is given, following [166], by the well-known formula [179]:

$$\Phi_{cm}(E_\nu) = b\, E_\nu^2\, E_e\, \sqrt{E_e^2 - m_e^2}\, F(\pm Z, E_e)\, \Theta(E_e - m_e) \qquad (5.1)$$

where the constant $b = \ln 2/m_e^5 ft_{1/2}$, with $m_e$ the electron mass and $ft_{1/2}$ the ft-value. The quantities appearing in the above expression are the energy $E_e = Q - E_\nu$ of the emitted lepton (electron or positron), $Q$ being the $Q$-value of the reaction, and the Fermi function $F(\pm Z, E_e)$, which accounts for the Coulomb modification of the spectrum.

In the laboratory frame, where the boosted nucleus has a velocity $v = \beta c$, the boosted flux $\Phi_{lab}(E_\nu, \theta)$ is given by [166]:

$$\Phi_{lab}(E_\nu, \theta) = \frac{\Phi_{cm}(E_\nu \gamma[1 - \beta \cos\theta])}{\gamma[1 - \beta \cos\theta]}\,. \qquad (5.2)$$

Considering a storage ring of total length $L$ with straight sections of length $S$, the total number of events per unit time in a cylindrical detector of radius $R$ and depth $h$, aligned with one of the straight sections of the storage ring, placed at a distance $d$ from the latter (position $C$ in Fig. 5.2) is:

$$\frac{dN_{ev}}{dt} = g\tau n h \times \int_0^\infty dE_\nu\, \Phi_{tot}(E_\nu)\, \sigma(E_\nu)\,, \qquad (5.3)$$

where $n$ is the number of target nuclei per unit volume, $\sigma(E_\nu)$ is the relevant neutrino-nucleus interaction cross section, and where

$$\Phi_{tot}(E_\nu) = \int_0^S \frac{d\ell}{L} \int_0^h \frac{dz}{h} \int_0^{\bar\theta(\ell,z)} \frac{\sin\theta d\theta}{2}\, \Phi_{lab}(E_\nu, \theta)\,, \qquad (5.4)$$

with

$$\tan \bar{\theta}(\ell, z) = \frac{R}{d + \ell + z}. \tag{5.5}$$

Following the discussion of [166] the neutrino fluxes are such that a "small" decay ring is far more efficient than a "large" decay ring in delivering neutrino events in the detector, the flux ratio (small ring/large ring) being proportional to the inverse ratio of the ring lengths $L_{SR}/L_{LR}$. The overall factor $L_{SR}/L_{LR}$ simply accounts for the fact that the number of decaying ions per unit length is smaller in a larger storage ring, and the solid angle covered by the most distant parts of the decay ring is much smaller.

It is to be noted that when the detector is placed close to the storage ring, the angular dependence of the neutrino flux detector is such that even inside the detector the fluxes significantly vary as functions of the transverse coordinate in the detector. This has led to extensive studies on the shape optimization of the close detector, as reported in [180]. This is in contrast with the case of a far detector considered in the high energy beta beam scenarios where the rate is practically insensitive to the transverse coordinate in the detector (see also Fig. 4.4).

Table 5.1, derived from [166], displays the number of events per year ($10^7$ s) collected for several reactions in a detector placed at $d = 10$ m away from a "small" ring ($L_{SR} = 450$ m, $S_{SR} = 150$ m) and large ring ($L_{LR} = 7$ km, $S_{LR} = 2.5$ km). From Table 5.1 it is evident that a small ring is needed to keep the number of events reasonably large and that it would result in an unpractical experiment collecting low energy events from the large ring required to accumulate ions at $\gamma = 100$ or higher, by putting a detector just in front of the straight section.

### 5.2.1   *Off-axis configurations*

The experimental possibility of running a low energy neutrino experiment close to the large ring needed by an oscillation experiment has been studied in [176] where the close detector is off-axis with respect to one of the decay ring straight sections. Most of the material in this section is taken from the above quoted article.

The accelerated ions emit the highest energy neutrinos along the boost direction. Therefore, by placing the detector off the storage ring straight section axis one gets rid of the highest energy component of the neutrino flux. The highest energy neutrinos reaching point $D$ in Fig. 5.2 will be emitted from the most distant point in the storage ring straight section

Table 5.1   Number of events per year for $\gamma = 7$ and $\gamma = 14$ in the small ($L_{SR} = 450$ m, $S_{SR} = 150$ m) and large ($L_{LR} = 7$ km, $S_{LR} = 2.5$ km) ring configurations. The detector is located at $d = 10$ m away from the ring and has dimensions $R = 1.5$ m and $h = 4.5$ m for the D ($D_2O$), $^{56}Fe$ and $^{208}Pb$, and $R = 4.5$ m and $h = 15$ m for the case of $^{16}O$ ($H_2O$), where $R$ is the radius and $h$ is the depth of the detector. The corresponding masses are given in tons. The flux-averaged cross section in the forward direction $\langle\sigma\rangle_\gamma$ in units of $10^{-42}$ cm$^2$ are also displayed. The relevant cross sections are taken from the indicated references. The results are obtained considering $2.2 \cdot 10^{18}$ $^{18}Ne$ decays/yr. From [166].

| Reaction | Ref. | Mass | $\gamma = 7$ | | | $\gamma = 14$ | | |
|---|---|---|---|---|---|---|---|---|
| | | | $\langle\sigma\rangle_\gamma$ | Small Ring | Large Ring | $\langle\sigma\rangle_\gamma$ | Small Ring | Large Ring |
| $\nu$+D | [181] | 35 | 36.3 | 65 | 5 | 184.57 | 784 | 60 |
| $\bar{\nu}$+D | [181] | 35 | 23.2 | 831 | 59 | 96.0 | 8668 | 652 |
| $\nu$+$^{16}O$ | [182] | 952 | 3.3 | 20 | 2 | 174.38 | 2018 | 245 |
| $\bar{\nu}$+$^{16}O$ | [182] | 952 | 5.0 | 708 | 64 | 102.0 | 27548 | 3151 |
| $\nu$+$^{56}Fe$ | [183] | 250 | 137.9 | 291 | 21 | 1402.1 | 6923 | 570 |
| $\nu$+$^{208}Pb$ | [184] | 360 | 2931.2 | 2533 | 182 | 16310.2 | 34632 | 2971 |

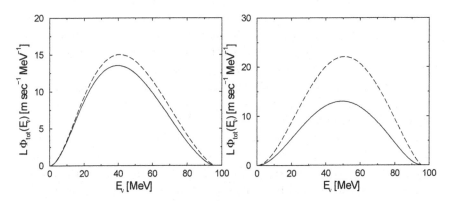

Fig. 5.3   Neutrino fluxes scaled by the length of the storage ring $L\Phi_{tot}(E_\nu)$: a small storage ring $SR$ (solid lines) and a large storage ring $LR$ (long dashed lines) are shown. The left (right) figure shows the fluxes impinging on the small (large) detector (see text). All fluxes are obtained with $^{18}Ne$ boosted at $\gamma = 14$. From [166]

(point $A$). If one wishes that only neutrinos with energy less or equal to $E_{cut}$ arrive at point $D$, the angle $\theta = \angle ADO$ has to satisfy the condition

$$\theta = \arccos\left[\frac{\gamma - (Q - m_e)/E_{cut}}{\sqrt{\gamma^2 - 1}}\right] . \qquad (5.6)$$

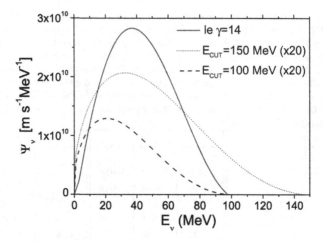

Fig. 5.4    Off-axis antineutrino fluxes ($\times 20$) evaluated at point $D$ of Fig. 5.2 for two different ion boosts and neutrino energy cuts. The neutrino flux from a low energy beta beam (le) at $\gamma = 14$ is shown for comparison. From [176].

The actual location of the detector depends on the desired antineutrino cut-off energy.

Figure 5.4 displays the off-axis antineutrino fluxes evaluated at point $D$, which lies on the perpendicular to the storage ring straight section derived from the turning point $O$ ($x = 0$; see Figure 5.2). The distance $y = AD = AO \cdot \tan \theta$ is determined using Eq. (5.6) and by constraining the maximum energy of the neutrinos ($E_{cut}$) reaching that point. The presented results correspond to the cases where the $^6$He ions are boosted at $\gamma = 100$, and when $E_{cut}$ is set to 100 and 150 MeV. In particular, for $\gamma = 100$ and $E_{cut} = 100$ MeV, the distance $y$ is 61.4 m. The main characteristics of such fluxes are summarized in Table 5.2. In order to compare these results with the low energy beta beam fluxes of Table 5.1, the values of $\widetilde{\Psi}_{\max}$ and $N_{ev}$ are normalized by the same detector volume. From Table 5.2 one can see that the off-axis antineutrino flux profiles are determined by the choice of $E_{cut}$ (which determines the angle $\theta$) and are not very sensitive to the boost of the ions. The flux shapes are strongly asymmetric, centered at low energies, and have rather long high energy tail.

From these results it is clear that both the off-axis flux at the peak intensity and the related number of events $N_{ev}$ are considerably smaller – by factors of 20 - 100 – than those of the low energy beta beam option. Such drastic reduction clearly makes this option hardly realizable for low energy neutrino physics applications, unless higher ion intensities are achieved.

Table 5.2 Average energy, peak energy evaluated at $\widetilde{\Psi}_{max}$ and peak-flux $(/10^9)$ for an off-axis flux for a cylindrical detector with $r = 4.5$ m and $h = 15$ m placed at $x = 0$ and $y$ (see Fig. 5.2) such that the maximum energy of the neutrinos is $E_{cut}$. $N_{ev}$ gives the number of events in $10^7$ s for the antineutrino scattering on protons considering water as a target material. The fluxes are in units of m MeV$^{-1}$s$^{-1}$. The ions are boosted at $\gamma = 100$; [176].

| $E_{cut}$ (MeV) | $\langle E \rangle$ $(MeV)$ | $E_{peak}$ (MeV) | $\widetilde{\Psi}_{max}$ | $N_{ev}$ | $y$ (m) |
|---|---|---|---|---|---|
| 100 | 29.3 | 18.5 | 0.57 | 135 | 61.4 |
| 150 | 43.6 | 28.0 | 0.87 | 666 | 48.0 |

In order to remove the high energy neutrinos from the flux, the off-axis detector should be placed relatively far away from the straight section ($y > 50$ m). This renders the intensities very low.

To overcome this difficulty, a setup made by two off-axis detectors has been proposed [176]. By using a subtraction procedure it is possible to extract a more intense low energy antineutrino flux compared to the single detector setup.

Considering the neutrino fluxes at two points $D_1$ and $D_2$, as shown in Figure 5.2, the neutrino flux $\widetilde{\Psi}_{D_1}(E_\nu)$ at point $D_1$ is split into two parts: one component produced in segment $AB$ of the storage ring, $\widetilde{\Psi}_{D_1}^{(AB)}(E_\nu)$, and the other produced in segment $BO$, $\widetilde{\Psi}_{D_1}^{(BO)}(E_\nu)$. The flux fraction $\widetilde{\Psi}_{D_1}^{(AB)}(E_\nu)$ at point $D_1$ is proportional to the neutrino flux $\widetilde{\Psi}_{D_2}(E_\nu)$ at point $D_2$ :

$$\widetilde{\Psi}_{D_1}^{(AB)}(E_\nu) = \widetilde{\Psi}_{D_2}(E_\nu)\frac{AO}{AB} . \tag{5.7}$$

The flux $\widetilde{\Psi}_{D_1}^{(BO)}(E_\nu)$ can be obtained by combining the responses of the two detectors located at $D_1$ and $D_2$:

$$\widetilde{\Psi}_{D_1}^{(BO)}(E_\nu) = \widetilde{\Psi}_{D_1}(E_\nu) - \widetilde{\Psi}_{D_2}(E_\nu)\frac{AO}{AB} . \tag{5.8}$$

Note that this flux contains only neutrinos with energies less than $E_{cut}$, set by Eq. (5.6). The subtracted flux of Eq. (5.8) has a similar energy dependence as the flux at the point

$$x = AD_1 \cos\theta - AO , \tag{5.9}$$

$$y = AD_1 \sin\theta , \tag{5.10}$$

but its intensity is higher by a factor of $y/y_{D_1}$. Here, the detector $D_1$ can be placed much closer to the storage ring with respect to the single detector setup where the $y$-distance is of 48 - 75 m. This implies a potential neutrino flux intensity enhancement by $\sim 10$ when $y_{D_1} = 5$ m.

The position of the detector at $D_2$ with respect to the position of the detector at $D_1$ is fixed by the choice of the desired maximal neutrino energy ($E_{cut}$) of the subtracted flux.

The flux gain is partially washed out when the realistic, large size detectors are considered where the flux varies very significantly as a function of the transverse coordinate.

Table 5.3    Flux characteristics for a two-detector setup ($y = 5$ m, see Fig. 5.2) computed for $^6$He at $\gamma = 100$. From [172].

| $E_{cut}$ (MeV) | $\langle E \rangle$ (MeV) | $E_{peak}$ (MeV) | $\widetilde{\Psi}_{max}$ ($\times 10^9$) |
|---|---|---|---|
| 100 | 18.5 | 6.5 | 2.42 |
| 100 | 28.2 | 10.8 | 3.05 |

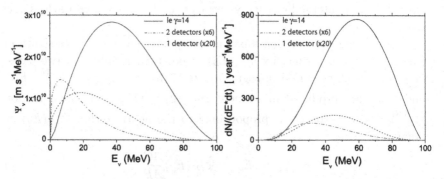

Fig. 5.5   Comparison of the different low energy neutrino fluxes (left panel) and the corresponding number of antineutrino-hydrogen events for the water detectors (right panel). The presented results are obtained for a standard $^6$He ($\gamma = 100$) beta beam exploiting two detectors off-axis (Fig. 5.2) and the subtraction method described in the text (these fluxes are multiplied by 6). As a comparison, the fluxes from a low energy beta beam (le), and for a single off-axis detector as described in Section II.B (HE), are given. The last flux is multiplied by 20. From [176].

Figure 5.5 shows the subtracted neutrino fluxes as well as differential number of events considering the same detector in the previous section, duplicated in $D_1$ and $D_2$ and aligned along $BD_1$ (see Fig. 5.2). The event rates are obtained by using the subtracted fluxes multiplied by the antineutrino on proton cross sections, considering that the detectors are filled with water. The subtracted flux characteristics are given in Table 5.3 for two different neutrino maximum energy cuts (100 and 150 MeV). The flux profile is even more asymmetric than for the single off-axis detector case (Fig. 5.4). Note that the average energy is pushed towards much lower energies (around 10 - 20 MeV) compared to the low energy beta beam flux. The expected intensities are significantly higher than in the case of the single off-axis detector, but still several times lower than the small ring on-axis detector setup.

## 5.3   Nuclear Structure, Neutrino-nucleus, Nuclear Astrophysics Applications

Neutrino-nucleus interactions represent a topic of current great interest for various domains of physics, from neutrino physics to nuclear physics and astrophysics. The motivations come for example from the need for a precise knowledge of the neutrino detector response in neutrino experiments and in core collapse supernova observatories aiming at the detection of the relic supernova neutrino background [185] using neutrino interaction on argon [186] and carbon or oxygen [187] or of neutrinos from an (extra)galactic explosion [185].

For instance, the 1n or 2n emission associated with charged-current events in a supernova lead-based observatory depends on the average electron neutrino energy, which encodes information on the still unknown third neutrino mixing angle $\theta_{13}$ [188]. Such a detector is now planned at SNO-LAB (the HALO project).

Neutrino-nucleon reactions play a crucial role in the understanding of the supernova dynamics [189, 190], the yields of the r-process nucleosynthesis that could take place in such environments [191] and also contribute to the energy transfer (from accretion-disk neutrinos to nucleons) in gamma-ray burst models [192, 193]. Finally, understanding the subtleties of the neutrino-nucleon interactions is crucial to the terrestrial observation of neutrino signals [194, 195].

Besides the astrophysical applications, a precise knowledge of the nuclear response of neutrinos is also crucial for our knowledge of the nuclear isospin and spin-isospin response that has fundamental implications, for example the search of physics beyond the Standard Model through neutrinoless double-beta decay [168].

In [165] it has been pointed out that the availability of neutrino beams in the 100 MeV energy range offers a unique opportunity to study spin-isospin and isospin nuclear excitations, for which little experimental information is available today. Both the isospin and spin-isospin collective (and non-collective) modes are excited when a neutrino encounters a nucleus, due to the vector and axial-vector nature of the weak interaction. The (super-) allowed Fermi transitions – due to the vector current and therefore the isospin operator – offer a well-known example of such excitations at low momentum transfer. A precise knowledge of these nuclear transitions is essential for determining the unitarity of the CKM matrix, the analog of the MNSP matrix in the quark sector (see e.g. [196]).

Another less known but still intriguing example is furnished by the allowed Gamow-Teller transitions – due to the axial-vector current and therefore the spin-isospin operator – in mirror nuclei, which are used to search for the possible existence of second-class currents in the weak interaction [197, 198]. These terms transform in the opposite way under the $\mathcal{G}$-parity transformation [1] as the usual vector and axial-vector terms, and are not present in the Standard Model. Because of their importance, both the allowed Fermi and Gamow-Teller transitions have been studied for a very long time in nuclear physics through beta-decay and charge-exchange reactions (for the Gamow-Teller ones). While the precision achieved in the description of the (super-)allowed Fermi transitions is impressive, our understanding of the allowed Gamow-Teller transitions still requires the use of an effective axial-vector coupling form factor, to take into account the "quenching" of the predicted transitions, compared to the ones measured in beta-decay or charge-exchange reactions (see for example [199]).

The still open "quenching" problem represents a limitation in our description of the weak spin-isospin nuclear response, in spite of the crucial role that it plays in various hot issues in nuclear astrophysics and high energy physics.

Little or no experimental information is available for the spin-isospin and isospin nuclear excitations such as the spin-dipole ($J^\pi = 0^-, 1^-, 2^-$), or the states of higher multipolarity (e.g. $J^\pi = 2^+, 3^-, 3^+, 4^-, 4^+$). These states come into play when a weak probe transfers a finite momentum to a nucleus, like in muon capture or in neutrino-nucleus interactions. Supplementary information on the corresponding weak transition amplitudes can be furnished by electron scattering studies, which however explore nuclear excitations induced by the vector current only.

---

[1] The $\mathcal{G}$-parity transformation corresponds to the product of the charge conjugation and of a rotation in isospin space.

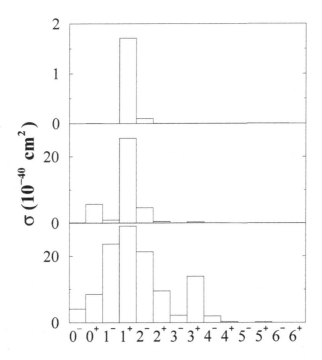

Fig. 5.6 Nuclear structure studies with low energy beta beams: contribution of isospin and spin-isospin nuclear states excited in the charged-current $^{208}$Pb$(\nu_e, e^-)^{208}$Bi reaction ($10^{-40}$ cm$^2$) for increasing neutrino energy, i.e. $E_{\nu_e} = 15$ MeV (up), 30 MeV (middle), 50 MeV (bottom). The histograms show the isobaric analogue state ($J^\pi = 0^+$), the allowed Gamow-Teller ($J^\pi = 1^+$), the spin-dipole ($J^\pi = 0^-, 1^-, 2^-$), as well as states of higher multipolarity ($J^\pi = 2^+, 3^-, 3^+, 4^-, 4^+$). For the latter no experimental information is available. Their contribution to the total cross section becomes significant when the impinging neutrino energy increases [165].

Figure 5.6 illustrates the contribution of spin-isospin and isospin transitions excited in neutrino scattering on lead, and their evolution when the neutrino energy increases.

A quantitative estimate of the importance of such states is also gathered by computing the flux-averaged cross sections[2], which are the relevant quantities for experiments. If one considers the neutrino fluxes corresponding to the decay-at-rest of muons, the spin-dipole states ($J^\pi = 0^-, 1^-, 2^-$) contribute by about 40% in $^{12}$C [200] and $^{56}$Fe [201], and by about 68% in $^{208}$Pb [184]. The contribution from the states of higher multipolarity is

---

[2]Flux-averaged cross sections are obtained by folding the cross sections with the neutrino flux of the source.

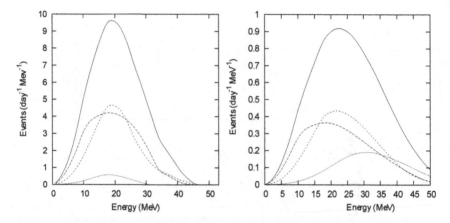

Fig. 5.7 Total electron spectrum coming from charged current neutrino scattering on lead (solid line), with neutrinos coming from pion decay-at-rest (left) and from low energy beta beams with $\gamma = 10$ (right). The other curves present the contribution from the allowed Fermi and Gamow-Teller states (long dashed line), from the $0^-, 1^-, 2^-$ (short dashed lines) and the $2^+, 3^-, 3^+, 4^-, 4^+$ states (dotted line) [167].

about 5% and 25% in iron and lead respectively; while it amounts to about 30% in carbon [200] and 60% in lead [184] if neutrinos are produced from pion decay-in-flight.

Since low energy beta beams have the specificity that the average neutrino energy can be raised by increasing the Lorentz boost of the ions, they constitute a promising tool for the study of these states, through a systematic study on various nuclear targets and different neutrino energies. Even though the measured cross sections are, in the majority of cases, inclusive, experimental information on these states can be extracted by changing the Lorentz ion boosts since different pieces of the nuclear response are important at different energies ([165], Fig. 5.6).

In [167] neutrino scattering on lead is taken as an example to show another procedure of extracting information on the different spin-isospin and isospin excitations, namely through a comparison of measurements with conventional beams and low energy beta beams (Fig. 5.7). In fact, the corresponding neutrino fluxes are in the same energy range for $\gamma = 7$, but their shape and average energies are different. Besides, the measurement of the cross section without or with (one or two) neutrons can be used for the same purpose, in the specific case of the lead nucleus [167].

A more extensive investigation as far as the nuclear structure information that can be extracted by performing measurements on several nuclei is done in [169]. In particular, the total charged-current and flux-averaged cross sections associated electron (anti)neutrino scattering on oxygen, iron, molybdenum and lead are investigated. The contribution from each multipole is given, showing that by using neutrinos from low energy beta beams, information on forbidden states, in particular the spin-dipole, can be extracted.

However, the available experimental neutrino-nucleus scattering data in the relevant energy range is limited, since deuteron, carbon and iron are the only nuclei investigated so far. As a consequence one has to rely on the numerous theoretical predictions and on extrapolations for the nuclei and energies of interest. These calculations exploit a variety of approaches, including Effective Field Theories, the Shell Model and the Random-Phase-Approximation, and the Elementary Particle Model [181, 222]. The cross section estimates agree quite well at very low energies, where the nuclear response is dominated by the allowed Fermi and Gamow-Teller transitions. However, important discrepancies appear at higher neutrino energies, when the other nuclear excitations become important (Figs. 5.6 and 5.7). In this energy region, the calculations are in fact largely subject to nuclear structure uncertainties and model dependencies (e.g. treatment of the continuum, choice of the forces, higher-order correlations).

The present discrepancies between the predicted and measured neutrino-carbon cross sections and between the calculations in the case of lead are talkative examples [223] of the difficult theoretical task. Systematic neutrino-nucleus interaction studies performed with low energy beta beams offer the perfect tool to explore the nuclear response in this energy region in great detail, and to put the theoretical predictions on firm ground.

A novel procedure to determine the response of a target nucleus in a supernova neutrino detector directly, through the use of low energy beta beams, is pointed out in [173, 175]. It is shown that the cross sections folded with a supernova neutrino spectrum can be well reproduced by linear combinations of beta beam spectra. This comparison offers a direct way to extract the main parameters of the supernova neutrino flux. The proposed procedure appears quite stable against uncertainties coming from the experiment, or the knowledge of the cross section, that give rise to "noise" in the expansion parameters.

Finally, it has been pointed out [168] that neutrino-nucleus interactions are also important in the search for neutrinoless double-beta decay in nuclei. In fact, by rewriting the neutrino exchange potential in momentum space and by using a multiple decomposition, the two-body transition operators, involved in the former, can be rewritten as a product of the one-body operators involved in neutrino-nucleus interactions (except for the short range correlations as well as possible phases present in the two-body process). Neutrino-nucleus scattering data offer a potential new constraint for the predictions on the neutrinoless double-beta decay half-lives. At present these calculations suffer from important discrepancies for the same candidate nucleus. Beta decay [202, 203], muon capture [204, 205], charge-exchange reactions [206, 207] and double-beta decay with the emission of two neutrinos [208] have been used to constrain the calculations so far. Neutrino-nucleus measurements would have the advantage that, if both neutrinos and antineutrinos are available, the nuclear matrix elements involved in the two branches of neutrinoless double-beta decay – from the initial and the final nucleus to the intermediate one – can be explored.

## 5.4 Fundamental Interaction Studies

Several applications for fundamental interaction studies of low energy beta beams have been discussed so far: the measurement of the Weinberg angle at low momentum transfer [171], a conserved vector current (CVC) test with neutrino beams [172], the measurement of the neutrino magnetic moment [167], the measurement of coherent neutrino-nucleus elastic scattering [210], the sensitivity to extra neutral gauge bosons, leptoquarks and r-parity breaking interactions [211].

### 5.4.1 *Weinberg angle measurement*

The measurement of the Weinberg angle represents an important test of the electroweak theory. Several experiments at different $Q^2$ exist, namely the atomic parity violation [225] and Moller scattering at $Q^2 = 0.026$ GeV$^2$ [226] which combined with the measurements of $\sin^2 \theta_W$ at the $Z^0$ pole [227], are consistent with the expected running of the weak mixing angle. However, recent measurement of the neutral- to charged-current ratio in muon antineutrino-nucleon scattering at the NuTEV experiment disagrees with these results by about 3 $\sigma$ [228]. A number of ideas were put forward to explain the so-called NuTEV anomaly [229–232]. However, a complete understanding of the physics behind it is still lacking; probing the Weinberg angle through additional experiments with different systematic errors would be very useful. The possibility of using a low energy beta-beam facility to

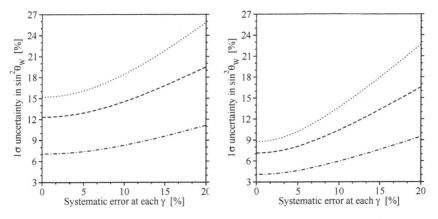

Fig. 5.8 One sigma uncertainty in the Weinberg angle as a function of the systematic error at each $\gamma$ for $\gamma = 12$ (dotted line), for $\gamma = 7, 12$ (broken line), and for $\gamma = 7, 8, 9, 10, 11, 12$ (dash-dotted line). The helium-6 intensity at the storage ring is $2.7 \times 10^{12}$ ions/s in the left panel [177], and with an increased intensity of $8.1 \times 10^{12}$ ions/s in the right panel. In both cases, the measurement duration at each $\gamma$ is $3 \times 10^7$ s [171].

carry out such a test with neutrino-electron scattering at low momentum transfer, i.e. $Q^2 = 10^{-4}$ GeV$^2$ is investigated in [171]. In particular it is shown that since the neutrino flux and average energy are well-known in the case of beta beams, the number of counts is in principle sufficient to extract information on the Weinberg angle. A fully efficient 1 kt Čerenkov detector is designed to be located 10 m from a small storage ring 1885 m long, 678 m straight sections, in which the expected intensities for $\gamma = 7\text{-}14$ are $0.5 \times 10^{11}$ helium-6/s and $2.7 \times 10^{12}$ neon-18/s. These numbers are the outcome of a preliminary feasibility study of the small storage ring [177]. The (anti)neutrino on electron events are identified by an angular cut. The background from neutrino-proton scattering is suppressed by the use of gadolinium [233]. Figure 5.8 shows the precision with which the Weinberg angle can be measured after the inclusion of both statistical (running $3 \times 10^7$ s at each gamma) and systematic errors. In particular, if the systematic error can be kept below 10%, a precision of 10% seems to be within reach at a beta-beam facility.

## 5.4.2 *Conserved vector current hypothesis*

The CVC hypothesis connects weak and electromagnetic hadronic currents. Several tests of CVC have been performed in the past, concerning, in particular, the vector form factor, through super-allowed nuclear beta decay studies (see e.g. [196]). Verifying that the CVC hypothesis correctly predicts tensor terms – often referred to as weak magnetism – is of fundamental importance [234]. So far, this contribution to the weak currents has been tested in an experiment involving the beta decay of Gamow-Teller transitions in mirror nuclei in the $A = 12$ triad [235–237].

A test based on neutrino-nucleon collisions at low momentum using low energy beta beams is proposed in [172]. This would have, in particular, the advantage that there is no uncertainty coming from nuclear structure calculations. In [172] the sensitivity to the weak magnetism term that can be achieved in the $\bar{\nu}_e + p \to e^+ + n$ reaction, both with the total number of events and with the angular distribution of the emitted positrons, is studied. For this purpose the same setup as for the measurement of the Weinberg angle is taken, and the interaction of anti-neutrinos on protons in a water Čerenkov detector is considered.

The results show that when systematic errors are taken into account, the angular distribution is a much better tool than the total number of events to extract information on the weak magnetism form factor. In particular, if those errors are kept below 5%, a one year measurement of the weak magnetism is possible at a 1 $\sigma$ level of 9%, if the ions in the storage ring are boosted to $\gamma = 12$.

An even better measurement is expected if the ions are boosted to $\gamma > 12$, because of the increasing importance of the weak magnetism contribution with the impinging neutrino energy. This way of probing the weak magnetism form factor at low momentum transfer constitutes a new test of the conserved vector current hypothesis.

### 5.4.3 *Neutrino magnetic moment*

It is interesting to note that one might use the ion decay at rest as an intense neutrino source, in order to explore neutrino properties that are still poorly known, such as the neutrino magnetic moment [170]. Direct measurements to achieve improved limits are precious, since the observation of a large magnetic moment points to physics beyond the Standard Model. Once produced, the ions are fired to a target inside a $4\pi$ detector.

The measurement of the (anti)neutrino interaction with the electrons, as a function of the electron recoil, is then used to set limits on the neutrino magnetic moment. Current bounds come from direct measurements using reactor neutrinos, and are in the range $\mu_\nu < 0.9 - 4 \times 10^{-10} \mu_B$ [238–242]. From solar neutrino-electron scattering at Super-Kamiokande a limit of

$\mu_\nu < 1.5 \times 10^{-10} \mu_B$ at 90% CL has been obtained [194]. Upper limits in the range $10^{-11}$- $10^{-12} \mu_B$ are also inferred from astrophysical and cosmological considerations [243], the exact values being model-dependent.

The prospects of using low energy beta beams to improve the direct bounds has been studied in comparison with reactor neutrinos and a very intense tritium source [170]. While the advantage of using beta beams is that the neutrino flux is perfectly known, the main limitation for this application is clearly the intensity of the ions. The conclusion of these studies is that sensitivities on neutrino magnetic moment of a few $10^{-11} \mu_B$ could be reached only if $^6$He sources were pushed to intensities of the order of $10^{15}$ ions/s.

# Bibliography

[1] B. T. Cleveland *et al.*, *Measurement of the solar electron neutrino flux with the Homestake chlorine detector*, Astrophys. J. **496** (1998) 505.

[2] K. S. Hirata *et al.* [KAMIOKANDE-II Collaboration], *Observation of B-8 Solar Neutrinos in the Kamiokande-II Detector*, Phys. Rev. Lett. **63** (1989) 16.

[3] P. Anselmann *et al.* [GALLEX Collaboration], *Solar neutrinos observed by GALLEX at Gran Sasso.*, Phys. Lett. B **285** (1992) 376. M. Altmann *et al.* [GNO Collaboration], *Complete results for five years of GNO solar neutrino observations*, Phys. Lett. B **616** (2005) 174.

[4] A. I. Abazov *et al.*, *First results from the Soviet-American gallium experiment*, Nucl. Phys. Proc. Suppl. **19** (1991) 84. J. N. Abdurashitov *et al.* [SAGE Collaboration], *Measurement of the solar neutrino capture rate by the Russian-American gallium solar neutrino experiment during one half of the 22-year cycle of solar activity*, J. Exp. Theor. Phys. **95** (2002) 181 [Zh. Eksp. Teor. Fiz. **122** (2002) 211].

[5] Q. R. Ahmad *et al.* [SNO Collaboration], *Direct evidence for neutrino flavor transformation from neutral-current interactions in the Sudbury Neutrino Observatory*, Phys. Rev. Lett. **89** (2002) 011301.

[6] K. Eguchi *et al.* [KamLAND Collaboration], *First results from KamLAND: Evidence for reactor anti-neutrino disappearance*, Phys. Rev. Lett. **90** (2003) 021802.

[7] Y. Fukuda *et al.* [Super-Kamiokande Collaboration], *Evidence for oscillation of atmospheric neutrinos*, Phys. Rev. Lett. **81** (1998) 1562.

[8] M. C. Sanchez *et al.* [Soudan 2 Collaboration], *Observation of atmospheric neutrino oscillations in Soudan 2*, Phys. Rev. D **68** (2003) 113004.

[9] M. Ambrosio *et al.* [MACRO Collaboration], *Atmospheric neutrino oscillations from upward throughgoing muon multiple scattering in MACRO*, Phys. Lett. B **566** (2003) 35.

[10] Y. Ashie *et al.* [Super-Kamiokande Collaboration], *Evidence for an oscillatory signature in atmospheric neutrino oscillation*, Phys. Rev. Lett. **93** (2004) 101801.

[11] M. Apollonio *et al.* [CHOOZ Collaboration], *Search for neutrino oscilla-*

*tions on a long base-line at the CHOOZ nuclear power station,* Eur. Phys. J. C **27** (2003) 331.

[12] E. Aliu *et al.* [K2K Collaboration], *Evidence for muon neutrino oscillation in an accelerator-based experiment,* Phys. Rev. Lett. **94** (2005) 081802.

[13] D. G. Michael *et al.* [MINOS Collaboration], *Observation of muon neutrino disappearance with the MINOS detectors and the NuMI neutrino beam,* Phys. Rev. Lett. **97** (2006) 191801.

[14] A. Aguilar *et al.* [LSND Collaboration], *Evidence for neutrino oscillations from the observation of anti-nu/e appearance in a anti-nu/mu beam,* Phys. Rev. D **64** (2001) 112007.

[15] B. Armbruster *et al.* [KARMEN Collaboration], *Upper limits for neutrino oscillations anti-nu/mu → anti-nu/e from muon decay at rest,* Phys. Rev. D **65** (2002) 112001.

[16] P. Astier *et al.* [NOMAD Collaboration], *Search for nu/mu → nu/e oscillations in the NOMAD experiment,* Phys. Lett. B **570** (2003) 19, [arXiv:hep-ex/0306037].

[17] A. A. Aguilar-Arevalo *et al.* [The MiniBooNE Collaboration], *A search for electron neutrino appearance at the delta $m^2$ $1 \, eV^2$ scale,* Phys. Rev. Lett. **98** (2007) 231801

[18] M. Maltoni and T. Schwetz, *Sterile neutrino oscillations after first Mini-BooNE results,* Phys. Rev. D **76** (2007) 093005 [arXiv:0705.0107 [hep-ph]].

[19] C. Giunti and M. Laveder, $\nu_e$ *Disappearance in MiniBooNE,* Phys. Rev. D **77** (2008) 093002 [arXiv:0707.4593 [hep-ph]].

[20] L. Wolfenstein, Phys. Rev. D17 (1978) 2369. S.P. Mikheev and A.Y. Smirnov, Nuovo Cim. C9 (1986) 17.

[21] G. L. Fogli, E. Lisi, A. Marrone and A. Palazzo, *Global analysis of three-flavor neutrino masses and mixings,* Prog. Part. Nucl. Phys. **57** (2006) 742 [arXiv:hep-ph/0506083].

[22] T. Schwetz, M. Tortola and J. W. F. Valle, *Three-flavour neutrino oscillation update,* New J. Phys. **10**, 113011 (2008) [arXiv:0808.2016 [hep-ph]].

[23] F. Feruglio, A. Strumia and F. Vissani, *Neutrino oscillations and signals in beta and 0nu 2beta experiments,* Nucl. Phys. B **637** (2002) 345 [Addendum-ibid. B **659** (2003) 359].

[24] M. Fukugita and T. Yanagida, *Baryogenesis without grand unification,* Phys. Lett. B **174** (1986) 45.

[25] A. De Rujula, M. B. Gavela and P. Hernandez, *Neutrino oscillation physics with a neutrino factory,* Nucl. Phys. B **547**, 21 (1999).

[26] A. Blondel *et al.*, *ECFA/CERN studies of a European neutrino factory complex,* CERN-2004-002, ECFA-04-230.

[27] Beams for European Neutrino Experiments (BENE): Midterm scientific report. By BENE Steering Group (A. Baldini *et al.*). Jan 2006.

[28] http://www.hep.ph.ic.ac.uk/iss/

[29] B. Richter, *Conventional beams or neutrino factories: The next generation of accelerator-based neutrino experiments,* arXiv:hep-ph/0008222.

[30] A. Guglielmi, M. Mezzetto, P. Migliozzi and F. Terranova, *Measurement of three-family neutrino mixing and search for CP violation,* arXiv:hep-

ph/0508034, published in D. Bettoni *et al.*, *The High intensity frontier*, Phys. Rept. **434** (2006) 47.

[31] A. Blondel, A. Cervera-Villanueva, A. Donini, P. Huber, M. Mezzetto and P. Strolin, *Future neutrino oscillation facilities*, Acta Phys. Polon. B **37** (2006) 2077.

[32] J. Burguet-Castell, M. B. Gavela, J. J. Gomez-Cadenas, P. Hernandez and O. Mena, *On the measurement of leptonic CP violation*, Nucl. Phys. B **608** (2001) 301.

[33] H. Minakata and H. Nunokawa, *Exploring neutrino mixing with low energy superbeams*, JHEP **0110** (2001) 001.

[34] G. L. Fogli and E. Lisi, *Tests of three-flavor mixing in long-baseline neutrino oscillation experiments*, Phys. Rev. D **54** (1996) 3667.

[35] V. Barger, D. Marfatia and K. Whisnant, *Breaking eight-fold degeneracies in neutrino CP violation, mixing, and mass hierarchy*, Phys. Rev. D **65** (2002) 073023.

[36] P. Huber, M. Lindner and W. Winter, *Superbeams versus neutrino factories*, Nucl. Phys. B **645**, 3 (2002) [arXiv:hep-ph/0204352].

[37] P. Huber, M. Maltoni and T. Schwetz, *Resolving parameter degeneracies in long-baseline experiments by atmospheric neutrino data*, Phys. Rev. D **71**, 053006 (2005) [arXiv:hep-ph/0501037]. R. Gandhi, P. Ghoshal, S. Goswami, P. Mehta and S. Uma Sankar, *Probing the nu mass hierarchy via atmospheric nu/mu + anti-nu/mu survival rates in Megaton water Cerenkov detectors,* arXiv:hep-ph/0506145.

[38] O. Yasuda, *New plots and parameter degeneracies in neutrino oscillations*, New J. Phys. **6**, 83 (2004) [arXiv:hep-ph/0405005].

[39] M. Ishitsuka, T. Kajita, H. Minakata and H. Nunokawa, *Resolving neutrino mass hierarchy and CP degeneracy by two identical detectors with different baselines*, Phys. Rev. D **72**, 033003 (2005) [arXiv:hep-ph/0504026].

[40] A. Donini, E. Fernandez-Martinez, P. Migliozzi, S. Rigolin and L. Scotto Lavina, *Study of the eightfold degeneracy with a standard beta-beam and a super-beam facility*, Nucl. Phys. B **710**, 402 (2005) [arXiv:hep-ph/0406132].

[41] H. W. Atherton *et al.*, *Precise measurements of particle production by 400 GeV/c protons on Beryllium targets*, CERN-80-07.

[42] G. Ambrosini *et al.* [NA56/SPY Collaboration], *Measurement of charged particle production from 450-GeV/c protons on beryllium*, Eur. Phys. J. C **10** (1999) 605.

[43] M. G. Catanesi *et al.* [HARP Collaboration], *The Harp Detector At The Cern Ps*, Nucl. Instrum. Meth. A **571**, 527 (2007).

[44] M. G. Catanesi *et al.* [HARP Collaboration], *Measurement of the production cross-section of positive pions in p Al collisions at 12.9-GeV/c*, Nucl. Phys. B **732**, 1 (2006).

[45] [HARP Collaboration], *Measurement of the production cross-section of positive pions in the collision of 8.9-GeV/c protons on beryllium*, Eur. Phys. J. C **52**, 29 (2007) [arXiv:hep-ex/0702024]

[46] M. H. Ahn *et al.* [K2K Collaboration], *Measurement of neutrino oscillation by the K2K experiment*, Phys. Rev. D **74**, 072003 (2006).

[47] The Fermilab NuMI Group, *NumI Facility Technical Design Report* Fermilab Report NuMI-346, 1998.

[48] P. Adamson *et al.* [MINOS Collaboration], *Measurement of Neutrino Oscillations with the MINOS Detectors in the NuMI Beam*, Phys. Rev. Lett. **101**, 131802 (2008) [arXiv:0806.2237 [hep-ex]].

[49] F. Arneodo *et al.* [ICARUS Collaboration], Nucl. Instrum. and Meth. A **461** (2001) 324; P. Aprili *et al.*, *The ICARUS experiment* CERN-SPSC/2002-27, SPSC-P-323.

[50] OPERA Collaboration, CERN-SPSC-P-318, LNGS-P25-00; H. Pessard [OPERA Collaboration], arXiv:hep-ex/0504033.M. Guler *et al.* [OPERA Collaboration], *OPERA: An appearance experiment to search for $\nu_\mu \rightarrow \nu_\tau$ oscillations in the CNGS beam. Experimental proposal*, CERN-SPSC-2000-028.

[51] G. Acquistapace *et al.*, *The CERN neutrino beam to Gran Sasso* CERN 98-02, INFN/AE-98/05 (1998); CERN-SL/99-034(DI), INFN/AE-99/05 Addendum.

[52] Report to the Fermilab Director by the Proton Committee, November 9, 2004,
http://www.fnal.gov/directorate/program_planning/Nov2004PACPublic/Draft_Proton_Plan_v2.pdf

[53] M. Komatsu, P. Migliozzi and F. Terranova, *Sensitivity to Theta(13) of the CERN to Gran Sasso neutrino beam*, J. Phys. G **29** (2003) 443.

[54] Y. Itow *et al.*, *The JHF-Kamiokande neutrino project*, arXiv:hep-ex/0106019.

[55] The E889 Collaboration, *Long Baseline Neutrino Oscillation Experiment at the AGS*, Brookhaven National Laboratory Report BNL No. 52459, April 1995. A. Para and M. Szleper, *Neutrino oscillations experiments using off-axis NuMI beam*, arXiv:hep-ex/0110032.

[56] K. Nishikawa, *Status of the T2K Experiment at J-Park*, Proceedings of *Neutrino Telescopes 2007*, 197-210, edited by Milla Baldo Ceolin, Venezia, 2007.

[57] D. S. Ayres *et al.* [NOvA Collaboration], *NOvA proposal to build a 30-kiloton off-axis detector to study neutrino oscillations in the Fermilab NuMI beamline*, arXiv:hep-ex/0503053.

[58] F. Ardellier *et al.* [Double Chooz Collaboration], *Double Chooz: A search for the neutrino mixing angle theta(13)*, arXiv:hep-ex/0606025.

[59] P. Huber, J. Kopp, M. Lindner, M. Rolinec and W. Winter, *From Double Chooz to Triple Chooz: Neutrino physics at the Chooz reactor complex*, JHEP **0605**, 072 (2006).

[60] X. Guo *et al.* [Daya Bay Collaboration], *A precision measurement of the neutrino mixing angle theta(13) using reactor antineutrinos at Daya Bay*, arXiv:hep-ex/0701029.

[61] P. Huber, M. Lindner, M. Rolinec, T. Schwetz and W. Winter, *Combined potential of future long-baseline and reactor experiments*, Nucl. Phys. Proc. Suppl. **145** (2005) 190.

[62] O. Mena, H. Nunokawa and S. J. Parke, *NOvA and T2K: The race for the neutrino mass hierarchy*, Phys. Rev. D **75** (2007) 033002.

[63] T. Kobayashi, *Super muon-neutrino beams: Physics reach and open questions*, J. Phys. G 29, 1493 (2003).

[64] P. Huber, M. Mezzetto and T. Schwetz, *On the impact of systematical uncertainties for the CP violation measurement in superbeam experiments*, JHEP **0803**, 021 (2008) [arXiv:0711.2950 [hep-ph]].

[65] T. Kajita, H. Minakata, S. Nakayama and H. Nunokawa, *Resolving eightfold neutrino parameter degeneracy by two identical detectors with different baselines*, Phys. Rev. D **75** (2007) 013006. M. Ishitsuka, T. Kajita, H. Minakata and H. Nunokawa, *Resolving neutrino mass hierarchy and CP degeneracy by two identical detectors with different baselines*, Phys. Rev. D **72** (2005) 033003.

[66] A. Meregaglia and A. Rubbia, *Neutrino oscillation physics at an upgraded CNGS with large next generation liquid argon TPC detectors*, JHEP **0611** (2006) 032.

[67] M. Meddahi, E. Shaposhnikova, *Analysis of the maximum potential proton flux to CNGS*, CERN-AB-2007-013.

[68] B. Baibussinov *et al.*, *A new, very massive modular Liquid Argon Imaging Chamber to detect low energy off-axis neutrinos from the CNGS beam. (Project MODULAr)*, Astropart. Phys. **29**, 174 (2008) [arXiv:0704.1422 [hep-ph]].

[69] B. Autin *et al.*, *Conceptual design of the SPL, a high-power superconducting H- linac at CERN*, CERN-2000-012.

[70] R. Garoby, *The SPL at CERN*, CERN-AB-2005-007.

[71] http://paf.web.cern.ch/paf/. O. Bruning *et al.*, *LHC luminosity and energy upgrade: A feasibility study*, CERN-LHC-PROJECT-REPORT-626; W. Scandale, Nucl. Phys. Proc. Suppl. **154** (2006) 101.

[72] J. J. Gomez-Cadenas *et al.*, *Physics potential of very intense conventional neutrino beams*, Proceedings of *Venice 2001, Neutrino telescopes* vol. 2\*, 463-481 [arXiv:hep-ph/0105297]. A. Blondel *et al.*, *Superbeam studies at CERN*, Nucl. Instrum. Meth. A **503** (2001) 173. M. Mezzetto, J. Phys. G **29** (2003) 1771.

[73] M. Apollonio *et al.*, *Oscillation physics with a neutrino factory*, arXiv:hep-ph/0210192.

[74] J. E. Campagne and A. Cazes, *The theta(13) and delta(CP) sensitivities of the SPL-Frejus project revisited*, Eur. Phys. J. C **45**, 643 (2006)

[75] M. Mezzetto, *SPL and beta beams to the Frejus*, Nucl. Phys. Proc. Suppl. **149** (2005) 179. J. E. Campagne, *The SPL-Frejus physics potential*, Nucl. Phys. Proc. Suppl. **155**, 185 (2006) [arXiv:hep-ex/0511013].

[76] J. E. Campagne, M. Maltoni, M. Mezzetto and T. Schwetz, *Physics potential of the CERN-MemphYS neutrino oscillation project*, JHEP **0704** (2007) 003 [arXiv:hep-ph/0603172].

[77] M. V. Diwan *et al.*, *Very long baseline neutrino oscillation experiments for precise measurements of mixing parameters and CP violating effects*, Phys. Rev. D **68** (2003) 012002 M. Diwan *et al.*, *Proposal for an experimental*

program in neutrino physics and proton decay in the homestake laboratory, arXiv:hep-ex/0608023.

[78] A. de Bellefon *et al.*, *MEMPHYS: A large scale water Cerenkov detector at Frejus*, arXiv:hep-ex/0607026.

[79] C. K. Jung [UNO Collaboration] *Feasibility of a next generation underground water Cherenkov detector: UNO*, arXiv:hep-ex/0005046.

[80] M. H. Ahn *et al.* [K2K Collaboration], *Search for electron neutrino appearance in a 250-km long-baseline experiment*, Phys. Rev. Lett. **93** (2004) 051801.

[81] J. Altegoer *et al.* [NOMAD Collaboration], *The NOMAD experiment at the CERN SPS*, Nucl. Instrum. Meth. A **404**, 96 (1998).

[82] P. Astier *et al.* [NOMAD Collaboration], *Prediction of neutrino fluxes in the NOMAD experiment*, Nucl. Instrum. Meth. A **515** (2003) 800 [arXiv:hep-ex/0306022].

[83] S. Geer, *Neutrino beams from muon storage rings: Characteristics and physics potential*, Phys. Rev. D **57** (1998) 6989 [Erratum-ibid. D **59** (1999) 039903], [hep-ph/9712290].

[84] *International design study of the neutrino factory*, http://www.hep.ph.ic.ac.uk/ids/

[85] P. Zucchelli, *A novel concept for a neutrino factory: the beta-beam*, Phys. Let. B, **532** (2002) 166-172.

[86] B. Autin, M. Benedikt, M. Grieser, S. Hancock, H. Haseroth, A. Jansson, U. Köster, M. Lindroos, S. Russenschuck and F. Wenander, *The acceleration and storage of radioactive ions for a neutrino factory*, CERN/PS 2002-078 (OP), Nufact Note 121, Proceedings of Nufact 02, London, UK, 2002, J. Phys. G: Nucl. Part. Phys. 29 (2003) 1785-1795.

[87] C. Albright, V. Barger, J. Beacom, S. Brice, J. J. Gomez-Cadenas, M. Goodman, D. Harris, P. Huber, A. Jansson, M. Lindner, O. Mena, P. Rapidis, K. Whisnant, W. Winter, *The Neutrino Factory, Muon Collider Collaboration, Neutrino Factory and Beta Beam Experiments and Developments*, C. Albright *et al*, in APS Joint Study Report on the Future of Neutrino Physics, FNAL-TM-2259, 2004; arXiv:physics/0411123.

[88] H. R. Ravn and B. W. Allardyce, *On-Line Mass Separators, in Treatise on Heavy-Ion Science*, Edt. D. A. Bromley, Plenum Press, New York, 1989, ISBN 0-306-42949-7.

[89] N. Thiollière, J. C. David, V. Blideanu, D. Doré, B. Rapp, D. Ridikas, *Optimization of 6He production using W or Ta converter surrounded by BeO target assembly*, Internal note CEA, DAPNIA-06-274 and EURISOL note, 03-25-2006-0004.

[90] M. Hass, D. Berkovits, T.Y. Hirsh, V. Kumar, M. Lewitowicz and F. de-Oliveira, *Light radioisotopes for nuclear astrophysics and neutrino physics*, J. Physics **G35**, 104042 (2008).

[91] E. Bouquerel, J. Lettry and T. Stora *BeO dual target prototype for 6He production - a preliminary note*, EURISOL internal note, http://eurisol.org.

[92] M. Loislet and S. Mitrofanov, *Alternative production scenarios for* $^6He$ *and* $^{18}Ne$, Oral presentation at the 6th Beta-beam Task Meeting, EURISOL,

19th November 2007, http://eurisol.org.

[93] C. Rubbia, A. Ferrari, Y. Kadi and V. Vlachoudis, *Beam cooling with ionisation losses,* Nucl. Instrum. Meth. A **568** (2006) 475 [arXiv:hep-ph/0602032].

[94] Y. Mori, *Development of FFAG accelerators and their applications for intense secondary particle production,* Nucl. Instrum. and Methods A, **562** (2006) 591.

[95] D. Neuffer, Fermi National Laboratory: Muon Collider and accelerator division document database: NFMCC-doc-516, beams-doc-2856 (2007).

[96] C. Reed, J. Nolen, V. Novick, J. Specht, and Y. Momozaki, *A Liquid Lithium Thin Film Stripper for RIA,* in the proceedings the Seventh International Conference on Radioactive Nuclear Beams, Cortina d'Ampezzo, Italy, 2006.

[97] P. Sortais, J. L. Bouly, J. C. Curdy, T. Lamy, P. Sole, T. Thuillier, J. L. Vieux-Rochaz, D. Voulot, *ECRIS development for stable and radioactive pulsed beams,* Rev. Sci. Instr. **75** (2004) 1610.

[98] L. Hermansson and D. Reistad, *Electron cooling at CELSIUS,* Nucl. Instrum. Meth. A **441** (2000) 140.

[99] E. Wilson, *An Introduction to Particle Accelerators,* Oxford University Press, Oxford, 2001, ISBN 0-19-850829-8.

[100] P. N. Ostroumov, *Development of a medium-energy superconducting heavy-ion linac,* Phys. Rev. ST Acc. Beams 5 (2002) 030101.

[101] The EURISOL report, APPENDIX D: Post-Accelerator & Mass Separator for EURISOL, Edt. M.-H. Moscatello, GANIL, Caen, France, 2003, EUROPEAN COMMISSION CONTRACT No. HPRI-CT-1999-500001.

[102] L. J.Laslet and L. Resegotti, Proceedings of the 6th International Conference on High-Energy Accelerators, Cambridge, USA, p.150.

[103] A. Lachaize, A. Tkatchenko, *The Rapid Cycling Synchrotron of the EURISOL Beta Beam facility,* EURISOL DS task note 12-25-2008-0012.

[104] C. Omet et al, *Charge change-induced beam losses under dynamic vacuum conditions in ring accelerators,* New J. Phys. **8** (2006) 284, doi:10.1088/1367-2630/8/11/284

[105] M. Magistris and M. Silari, *Parameters of radiological interest for a beta-beam decay ring,* CERN technical note, CERN-TIS-2003-017-RP-TN.

[106] M. Kirk, *PS activation and collimation studies,* Oral presentation at the 6th Beta-beam Task Meeting, EURISOL, 19th November 2007, http://eurisol.org

[107] A. Chancé and J. Payet, *Beta-beam decay ring design,* In the proceedings of EPAC, 2006, Edinburgh, UK, p.1933.

[108] A. Chancé and J. Payet, *Transport of the decay products in the beta-beam decay ring,* In the proceedings of EPAC, 2008, Genoa, Italy. p.3104

[109] M. Benedikt and S. Hancock, *A novel scheme for injection and stacking of radioactive ions at high energy,* Nucl. Instrum. Meth. A **550** (2005) 1.

[110] F.W. Jones and E. Wildner, *Simulation of decays and secondary ion losses in a betabeam decay ring,* in the proceedings of PAC07, Albuquerque, New Mexico, USA

[111] R. Gupta, M. Anerella, M. Harrison, J. Schmalzle and N. Mokhov, *Open midplane dipole design for LHC IR upgrade,* IEEE Trans. Appl. Supercond. **14** (2004) 259.

[112] J. MacLachlan, *Particle Tracking in E-Phi Space for Synchrotron Design & Diagnosis,* Fermilab-CONF-92/333 (Nov. 92), presented at 12-th Int'l Conf. on Appl. of Acc. in Res. and Ind., Denton TX, 4 Nov 1992 and http://www-ap.fnal.gov/ESME/

[113] S. Hancock, M. Lindroos, E. McIntosh and M. Metcalf, *Tomographic measurements of longitudinal phase space density,* Comput. Phys. Commun. **118** (1999) 61.

[114] S. Hancock, *Technical challenges of the EURISOL beta-beam,* AIP Conf. Proc. **981** (2008) 89.

[115] http://snowmass2001.org/

[116] A. Källberg and M. Lindroos, *Accumulation in a ring at low energy for the beta-beam,* EURISOL DS/TASK12/TN-05-04, http://eurisol.org and In the proceedings of European Particle Accelerator Conference, 2006, Edinburgh, Scotland.

[117] M. Blaskiewicz and J. M. Brennan, *A Barrier Bucket Experiment for Accumulating De-bunched Beam in the AGS,* In the proceedings of the European Particle Accelerator Conference, Barcelona, Spain, 1996.

[118] J. Bernabeu, J. Burguet-Castell, C. Espinoza and M. Lindroos, *Monochromatic neutrino beams,* JHEP **0512**, 014 (2005) [arXiv:hep-ph/0505054].

[119] J. Sato, *Monoenergetic Neutrino Beam for Long-Baseline Experiments,* Phys. Rev. Lett. **95** (2005) 131804.

[120] E. Nacher, *Beta decay studies in the $N \sim Z$ and the rare-earth regions using Total Absorption Spectroscopy techniques,* Ph. D. Thesis, Univ. Valencia (2004).

[121] A. Fukumi, I. Nakano, H. Nanjo, N. Sasao, S. Sato, M. Yoshimura, *CP-even neutrino beam,* arXiv:hep-ex/0612047.

[122] M. Lindroos, J. Bernabeu, J. Burguet-Castell and C. Espinoza, *A monochromatic electron neutrino beam,* In the proceedings of International Europhysics Conference on High energy Physics, Lisboa, Portugal, 2005, Proceedings of Science, http://pos.sissa.it/

[123] F. Bosch, GSI, Germany, 2005, private communication.

[124] W. Bambynek, H. Behrens, M. H. Chen, B. Crasemann, M. L. Fitzpatrick, K. W. D. Ledingham, H. Genz, M. Mutterer and R. L. Intemann, *Orbital electron capture by the nucleus,* Rev. Mod. Phys. **49** (1977) 77.

[125] J. Burguet-Castell, D. Casper, J. J. Gomez-Cadenas, P. Hernandez and F. Sanchez, *Neutrino oscillation physics with a higher gamma beta-beam,* Nucl. Phys. B **695** (2004) 217 [arXiv:hep-ph/0312068].

[126] http://www.ganil.fr/eurisol/

[127] See for instance S. S. Masood, S. Nasri, J. Schechter, M. A. Tortola, J. W. F. Valle and C. Weinheimer, *Exact relativistic beta decay endpoint spectrum,* Phys. Rev. C **76** (2007) 045501 [arXiv:0706.0897 [hep-ph]].

[128] M. Mezzetto, *Physics reach of the beta beam,* J. Phys. G **29** (2003) 1771 [arXiv:hep-ex/0302007].

[129] J. Bouchez, M. Lindroos and M. Mezzetto, *Beta beams: Present design and expected performances*, AIP Conf. Proc. **721** (2004) 37 [arXiv:hep-ex/0310059].

[130] M. Mezzetto, *Physics potential of the gamma = 100,100 beta beam*, Nucl. Phys. Proc. Suppl. **155** (2006) 214

[131] D. Casper, *The nuance neutrino physics simulation, and the future*, Nucl. Phys. Proc. Suppl. **112** (2002) 161 [arXiv:hep-ph/0208030].

[132] H. Maesaka, *Evidence for muon neutrino oscillation in an accelerator-based experiment*, Ph. D thesis, Kyoto University, 2005.

[133] J. J. Aguilar-Arevalo *et al.* [SciBooNE coll.], *Bringing the SciBar detector to the booster neutrino beam*, hep-ex/0601022.

[134] D. Drakoulakos *et al.* [Minerva coll.], *Proposal to perform a high-statistics neutrino scattering experiment using a fine-grained detector in the NuMI beam*, hep-ex/0405002.

[135] P. Huber, M. Lindner and W. Winter, *Simulation of long-baseline neutrino oscillation experiments with GLoBES*, Comput. Phys. Commun. **167** (2005) 195. P. Huber, J. Kopp, M. Lindner, M. Rolinec and W. Winter, *New features in the simulation of neutrino oscillation experiments with GLoBES 3.0*, Comput. Phys. Commun. **177**, 432 (2007) [arXiv:hep-ph/0701187].

[136] S. T. Petcov, *Diffractive-like (or parametric-resonance-like?) enhancement of the earth (day-night) effect for solar neutrinos crossing the earth core*, Phys. Lett. B **434** (1998) 321 [hep-ph/9805262]; M. Chizhov, M. Maris and S. T. Petcov, *On the oscillation length resonance in the transitions of solar and atmospheric neutrinos crossing the earth core*, hep-ph/9810501; M. V. Chizhov and S. T. Petcov, Phys. Rev. Lett. **83** (1999) 1096 [hep-ph/9903399].

[137] E. K. Akhmedov, *Parametric resonance of neutrino oscillations and passage of solar and atmospheric neutrinos through the earth*, Nucl. Phys. B **538**, 25 (1999) [hep-ph/9805272]; E. K. Akhmedov, A. Dighe, P. Lipari and A. Y. Smirnov, *Atmospheric neutrinos at Super-Kamiokande and parametric resonance in neutrino oscillations*, Nucl. Phys. B **542**, 3 (1999) [hep-ph/9808270].

[138] J. Bernabeu, S. Palomares-Ruiz and S. T. Petcov, *Atmospheric neutrino oscillations, theta(13) and neutrino mass hierarchy*, Nucl. Phys. B **669**, 255 (2003) [hep-ph/0305152].

[139] C. W. Kim and U. W. Lee, *Comment on the possible electron-neutrino excess in the Super-Kamiokande atmospheric neutrino experiment*, Phys. Lett. B **444**, 204 (1998) [hep-ph/9809491].

[140] O. L. G. Peres, A. Y. Smirnov, *Atmospheric neutrinos: LMA oscillations, U(e3) induced interference and CP-violation*, Nucl. Phys. B **680** (2004) 479 [hep-ph/0309312].

[141] M. C. Gonzalez-Garcia, M. Maltoni, A.Y. Smirnov, *Measuring the deviation of the 2-3 lepton mixing from maximal with atmospheric neutrinos*, Phys. Rev. D **70** (2004) 093005 [hep-ph/0408170].

[142] T. Kajita, Talk at NNN05, 7–9 April 2005, Aussois, Savoie, France, http://nnn05.in2p3.fr/

[143]  M. C. Gonzalez-Garcia and M. Maltoni, *Atmospheric neutrino oscillations and new physics*, Phys. Rev. D **70** (2004) 033010 [hep-ph/0404085].

[144]  J. Burguet-Castell, D. Casper, E. Couce, J. J. Gomez-Cadenas and P. Hernandez, *Optimal beta-beam at the CERN-SPS*, Nucl. Phys. B **725**, 306 (2005)

[145]  P. Huber *et al.*, *Physics and optimization of beta-beams: From low to very high gamma*, Phys. Rev. D **73**, 053002 (2006).

[146]  F. Terranova, A. Marotta, P. Migliozzi and M. Spinetti, *High energy beta beams without massive detectors*, Eur. Phys. J. C **38** (2004) 69. A. Donini, E. Fernandez-Martinez, P. Migliozzi, S. Rigolin, L. Scotto Lavina, T. Tabarelli de Fatis and F. Terranova, *A beta beam complex based on the machine upgrades of the LHC*, Eur. Phys. J. C **48**, 787 (2006)

[147]  J. Bernabeu and C. Espinoza, *Energy Dependence of CP-Violation Reach for Monochromatic Neutrino Beam*, Phys. Lett. B **664**, 285 (2008) [arXiv:0712.1034 [hep-ph]].

[148]  M. Rolinec and J. Sato, *Neutrino beams from electron capture at high gamma*, JHEP **0708**, 079 (2007) [arXiv:hep-ph/0612148].

[149]  J. N. Bahcall, *Theory of Bound-State Beta Decay*, Phys. Rev. **194**, 495 (1961).

[150]  C. Rubbia, *Ionization cooled ultra pure beta-beams for long distance nu/e −− > nu/mu transitions, theta(13) phase and CP-violation*, arXiv:hep-ph/0609235.

[151]  A. Donini and E. Fernandez-Martinez, *Alternating ions in a beta-beam to solve degeneracies*, Phys. Lett. B **641**, 432 (2006) [arXiv:hep-ph/0603261].

[152]  D. Meloni, O. Mena, C. Orme, S. Palomares-Ruiz and S. Pascoli, *An intermediate gamma beta-beam neutrino experiment with long baseline*, JHEP **0807**, 115 (2008) [arXiv:0802.0255 [hep-ph]].

[153]  P. Huber and W. Winter, *Neutrino factories and the 'magic' baseline*, Phys. Rev. D **68**, 037301 (2003). A. Y. Smirnov, *Neutrino oscillations: What is magic about the 'magic' baseline?*, arXiv:hep-ph/0610198. H. Minakata, *Measuring earth matter density and testing the MSW theory*, arXiv:0705.1009 [hep-ph].

[154]  A. M. Dziewonski and D. L. Anderson, *Preliminary Reference Earth Model*, Phys. Earth Planet. Interiors **25**, 297 (1981).

[155]  S. K. Agarwalla, S. Choubey, S. Goswami and A. Raychaudhuri, *Neutrino parameters from matter effects in $P_{ee}$ at long baselines*, Phys. Rev. D **75** (2007) 097302 [arXiv:hep-ph/0611233]. S. K. Agarwalla, S. Choubey and A. Raychaudhuri, *Neutrino mass hierarchy and theta(13) with a magic baseline beta-beam experiment*, Nucl. Phys. B **771**, 1 (2007) [arXiv:hep-ph/0610333]. S. K. Agarwalla, S. Choubey and A. Raychaudhuri, *Unraveling neutrino parameters with a magical beta-beam experiment at INO*, Nucl. Phys. B **798**, 124 (2008) [arXiv:0711.1459 [hep-ph]].

[156]  See http://www.imsc.res.in/~ino.

[157]  P. Coloma, A. Donini, E. Fernandez-Martinez and J. Lopez-Pavon, $\theta_{13}$, $\delta$ and the neutrino mass hierarchy at a $\gamma = 350$ double baseline Li/B $\beta$-Beam, JHEP **0805**, 050 (2008) [arXiv:0712.0796 [hep-ph]].

[158] S. K. Agarwalla, S. Choubey, A. Raychaudhuri and W. Winter, *Optimizing the greenfield Beta-beam*, JHEP **0806** (2008) 090 [arXiv:0802.3621 [hep-ex]].

[159] M. Mezzetto, *Beta beams*, Nucl. Phys. Proc. Suppl. **143** (2005) 309.

[160] S. K. Agarwalla, S. Choubey and A. Raychaudhuri, *Exceptional Sensitivity to Neutrino Parameters with a Two Baseline Beta-Beam Set-up*, arXiv:0804.3007 [hep-ph].

[161] W. Winter, *Minimal Neutrino Beta Beam for Large $\theta_{13}$*, arXiv:0804.4000 [hep-ph].

[162] V. Barger, P. Huber, D. Marfatia and W. Winter, *Which long-baseline neutrino experiments are preferable?*, Phys. Rev. D **76** (2007) 053005 [arXiv:hep-ph/0703029].

[163] P. Huber and W. Winter, *Neutrino Factory Superbeam*, Phys. Lett. B **655** (2007) 251 [arXiv:0706.2862 [hep-ph]].

[164] Courtesy of Dr. Sanjib Agarwalla, Harish-Chandra Research Institute, Allahabad 211 019, India.

[165] C. Volpe, *What about a beta beam facility for low energy neutrinos?*, J. Phys. G **30** (2004) L1 [arXiv:hep-ph/0303222].

[166] J. Serreau and C. Volpe, *Neutrino nucleus interaction rates at a low-energy beta-beam facility*, Phys. Rev. C **70** (2004) 055502 [arXiv:hep-ph/0403293].

[167] G. C. McLaughlin, *Neutrino nucleus cross section measurements using stopped pions and low energy beta beams*, Phys. Rev. C **70** (2004) 045804 [arXiv:nucl-th/0404002].

[168] C. Volpe, *Neutrino nucleus interactions as a probe to constrain double-beta decay predictions*, J. Phys. G **31** (2005) 903 [arXiv:hep-ph/0501233].

[169] R. Lazauskas and C. Volpe, *Neutrino beams as a probe of the nuclear isospin and spin-isospin excitations*, Nucl.Phys.A792:219-228,2007, [arXiv:0704.2724] .

[170] G. C. McLaughlin and C. Volpe, *Prospects for detecting a neutrino magnetic moment with a tritium source and beta beams*, Phys. Lett. B **591** (2004) 229 [arXiv:hep-ph/0312156].

[171] A. B. Balantekin, J. H. de Jesus and C. Volpe, *Electroweak tests at beta-beams*, Phys. Lett. B **634** (2006) 180 [arXiv:hep-ph/0512310].

[172] A. B. Balantekin, J. H. de Jesus, R. Lazauskas and C. Volpe, *A conserved vector current test using low energy beta-beams*, Phys. Rev. D **73** (2006) 073011 [arXiv:hep-ph/0603078].

[173] N. Jachowicz and G. C. McLaughlin, *Reconstructing supernova-neutrino spectra using low-energy beta-beams*, Phys. Rev. Lett. **96** (2006) 172301 [arXiv:nucl-th/0604046].

[174] C. Volpe, *Topical review on 'beta-beams'*, J. Phys. G **34** (2007) R1 [arXiv:hep-ph/0605033].

[175] N. Jachowicz, G. C. McLaughlin and C. Volpe, *Untangling supernova-neutrino oscillations with beta-beam data*, Phys.Rev.C77:055501,2008, arXiv:0804.0360 [nucl-th].

[176] R. Lazauskas, A. B. Balantekin, J. H. De Jesus and C. Volpe, *Low-energy*

neutrinos at off-axis from a standard beta-beam, Phys. Rev. D **76** (2007) 053006 [arXiv:hep-ph/0703063].

[177]  A. Chancé and J. Payet (2005) private communication

[178]  M. Benedikt, A. Chancé, M. Lindroos, J. Payet (2006) private communication

[179]  K. S. Krane, *Introductory Nuclear Physics*, John Wiley and Sons (1998).

[180]  P. S. Amanik and G. C. McLaughlin, *Manipulating a neutrino spectrum to maximize the physics potential from a low energy beta beam*, Phys. Rev. C **75**, 065502 (2007) [arXiv:hep-ph/0702207].

[181]  K. Kubodera and S. Nozawa, *Neutrino-nucleus reactions*, Int. J. Mod. Phys. E **3** (1994) 101 [arXiv:nucl-th/9310014].

[182]  E. Kolbe, K. Langanke and P. Vogel, *Estimates of weak and electromagnetic nuclear decay signatures for neutrino reactions in Super-Kamiokande*, Phys. Rev. D **66**, 013007 (2002). W. C. Haxton, *The nuclear response of water cherenkov detectors to supernova and solar neutrinos*, Phys. Rev. D **36**, 2283 (1987).

[183]  E. Kolbe and K. Langanke, *The role of nu induced reactions on lead and iron in neutrino detectors*, Phys. Rev. C **63**, 025802 (2001) [arXiv:nucl-th/0003060].

[184]  C. Volpe, N. Auerbach, G. Colo and N. Van Giai, *Charged-current neutrino Pb-208 reactions*, Phys. Rev. C **65**, 044603 (2002) [arXiv:nucl-th/0103039].

[185]  D. Autiero *et al.*, *Large underground, liquid based detectors for astro-particle physics in Europe: Scientific case and prospects*, JCAP 0711:011,2007 [arXiv:0705.0116].

[186]  A. G. Cocco, A. Ereditato, G. Fiorillo, G. Mangano and V. Pettorino, *Supernova relic neutrinos in liquid argon detectors*, JCAP **0412** (2004) 002 [arXiv:hep-ph/0408031].

[187]  C. Volpe, J. Welzel, *Supernova Relic Electron Neutrinos and anti-Neutrinos in future Large-scale Observatories.*, arXiv:0711.3237.

[188]  J. Engel, G. C. McLaughlin and C. Volpe, *What can be learned with a lead based supernova neutrino detector?* Phys. Rev. D **67** (2003) 013005 [arXiv:hep-ph/0209267].

[189]  A. B. Balantekin and G. M. Fuller, *Supernova neutrino nucleus astrophysics*, J. Phys. G **29**, 2513 (2003) [arXiv:astro-ph/0309519].

[190]  C. J. Horowitz, *Weak magnetism for antineutrinos in supernovae*, Phys. Rev. D **65**, 043001 (2002) [arXiv:astro-ph/0109209].

[191]  B. S. Meyer, G. C. McLaughlin and G. M. Fuller, *Neutrino capture and r-process nucleosynthesis*, Phys. Rev. C **58**, 3696 (1998) [arXiv:astro-ph/9809242].

[192]  M. Ruffert, H. T. Janka, K. Takahashi and G. Schaefer, *Coalescing neutron stars − − > a step towards physical models. II. Neutrino emission, neutron tori, and gamma-ray bursts*, Astron. Astrophys. **319** (1997) 122 [arXiv:astro-ph/9606181].

[193]  J. P. Kneller, G. C. McLaughlin and R. Surman, *Neutrino scattering, absorption and annihilation above the accretion disks of gamma ray bursts*, J. Phys. G **32**, 443 (2006) [arXiv:astro-ph/0410397].

[194] P. Vogel and J. F. Beacom, *The angular distribution of the neutron inverse beta decay, anti-nu/e + p -- > e$^+$ + n*, Phys. Rev. D **60**, 053003 (1999) [arXiv:hep-ph/9903554].

[195] J. F. Beacom, W. M. Farr and P. Vogel, *Detection of supernova neutrinos by neutrino proton elastic scattering*, Phys. Rev. D **66** (2002) 033001 [arXiv:hep-ph/0205220].

[196] J. C. Hardy and I. S. Towner, *Superallowed 0$^+$ - 0$^+$ nuclear beta decays: A critical survey with tests of CVC and the standard model*, Phys. Rev. C **71**, 055501 (2005) [arXiv:nucl-th/0412056].

[197] S. Weinberg, *Charge symmetry of weak interactions*, Phys. Rev. **112**, 1375 (1958).

[198] D. H. Wilkinson, *Limits to second-class nucleonic and mesonic currents*, Eur. Phys. J. A **7**, 307 (2000).

[199] F. Osterfeld, *Nuclear spin and isospin excitations*, Rev. Mod. Phys. **64**, 491 (1992).

[200] C. Volpe, N. Auerbach, G. Colo, T. Suzuki and N. Van Giai, *Microscopic theories of neutrino C-12 reactions*, Phys. Rev. C **62**, 015501 (2000) [arXiv:nucl-th/0001050].

[201] E. Kolbe, K. Langanke and G. Martinez-Pinedo, *The inclusive $^{56}Fe(nue,e^-)^{56}Co$ cross section*, Phys. Rev. C **60**, 052801 (1999) [arXiv:nucl-th/9905001].

[202] K. Muto, E. Bender and H. V. Klapdor, *Effects of ground state correlations on 2 neutrino beta beta decay rates and limitations of the QRPA approach*, Z. Phys. A **334**, 177 (1989).

[203] M. Aunola and J. Suhonen, *Systematic study of beta and double beta decay to excited final states*, Nucl. Phys. A **602**, 133 (1996).

[204] M. Kortelainen and J. Suhonen, *Ordinary muon capture as a probe of virtual transitions of beta beta decay*, Europhys. Lett. **58**, 666 (2002) [arXiv:nucl-th/0201007].

[205] M. Kortelainen and J. Suhonen, *Nuclear muon capture as a powerful probe of double-beta decays in light nuclei*, J. Phys. G **30**, 2003 (2004).

[206] H. Akimune *et al.*, *GT strengths studied by (He-3,t) reactions and nuclear matrix elements for double beta decays*, Phys. Lett. B **394**, 23 (1997) [Erratum-ibid. B **665**, 424 (2008)].

[207] J. Bernabeu *et al* IPNO/TH-88-58, FTUV-88/20 (1998).

[208] V. A. Rodin, A. Faessler, F. Simkovic and P. Vogel, *On the uncertainty in the 0nu beta beta decay nuclear matrix elements*, Phys. Rev. C **68**, 044302 (2003) [arXiv:nucl-th/0305005].

[209] C. Volpe, *Physics potential of beta-beams*, Nucl. Phys. A **752** (2005) 38 [arXiv:hep-ph/0409357].

[210] A. Bueno, M. C. Carmona, J. Lozano and S. Navas, *Observation of coherent neutrino-nucleus elastic scattering at a beta beam*, Phys. Rev. D **74**, 033010 (2006).

[211] J. Barranco, O. G. Miranda and T. I. Rashba, *Low energy neutrino experiments sensitivity to physics beyond the standard model*, Phys. Rev. D **76**, 073008 (2007) [arXiv:hep-ph/0702175].

[212] C. Volpe, *Physics with a very first low-energy beta-beam*, Nucl. Phys. Proc. Suppl. **155** (2006) 97 [arXiv:hep-ph/0510242].

[213] N. A. Smirnova and C. Volpe, *On the asymmetry of Gamow-Teller beta-decay rates in mirror nuclei in relation with second-class currents*, Nucl. Phys. A **714** (2003) 441 [arXiv:nucl-th/0207078].

[214] G. C. McLaughlin and G. M. Fuller, *Neutrino capture on heavy nuclei*, Astrophys. J. **455**, 202 (1995).

[215] W. C. Haxton, K. Langanke, Y. Z. Qian and P. Vogel, *Neutrino-induced nucleosynthesis and the site of the r-process*, Phys. Rev. Lett. **78** (1997) 2694 [arXiv:astro-ph/9612047].

[216] Y. Z. Qian, W. C. Haxton, K. Langanke and P. Vogel, *Neutrino-induced neutron spallation and supernova r-process nucleosynthesis*, Phys. Rev. C **55**, 1532 (1997) [arXiv:nucl-th/9611010].

[217] I. N. Borzov and S. Goriely, *Weak interaction rates of neutron rich nuclei and the R process nucleosynthesis*, Phys. Rev. C **62** (2000) 035501.

[218] M. Terasawa, T. Kajino, K. Langanke, G. J. Mathews and E. Kolbe, *Neutrino effects before, during, and after the freezeout of the r-process*, Astrophys. J. **608**, 470 (2004).

[219] S. E. Woosley, D. H. Hartmann, R. D. Hoffman and W. C. Haxton, *The Neutrino Process*, Astrophys. J. **356**, 272 (1990).

[220] A. Heger, E. Kolbe, W. C. Haxton, K. Langanke, G. Martinez-Pinedo and S. E. Woosley, *Neutrino nucleosynthesis*, Phys. Lett. B **606**, 258 (2005) [arXiv:astro-ph/0307546].

[221] K. Langanke and G. Martinez-Pinedo, *Supernova Neutrino Nucleus Reactions*, Nucl. Phys. A **731**, 365 (2004).

[222] E. Kolbe, K. Langanke, G. Martinez-Pinedo and P. Vogel, *Neutrino nucleus reactions and nuclear structure*, J. Phys. G **29** (2003) 2569 [arXiv:nucl-th/0311022].

[223] C. Volpe, *Neutrino nucleus interactions: Open questions and future projects*, Nucl. Phys. Proc. Suppl. **143** (2005) 43 [arXiv:hep-ph/0409249], and references therein.

[224] M. Sajjad Athar, S. Ahmad and S. K. Singh, *Neutrino nucleus cross sections for low energy neutrinos at SNS facilities*, Nucl. Phys. A **764**, 551 (2006) [arXiv:nucl-th/0506046].

[225] S. C. Bennett and C. E. Wieman, *Measurement of the 6S – 7S transition polarizability in atomic cesium and an improved test of the standard model*, Phys. Rev. Lett. **82**, 2484 (1999) [arXiv:hep-ex/9903022].

[226] P. L. Anthony *et al.* [SLAC E158 Collaboration], *Precision measurement of the weak mixing angle in Moeller scattering*, Phys. Rev. Lett. **95** (2005) 081601 [arXiv:hep-ex/0504049].

[227] ALEPH, DELPHI, L3, OPAL, and SLD Collaborations 2005

[228] G. P. Zeller *et al.* [NuTeV Collaboration], *A precise determination of electroweak parameters in neutrino nucleon scattering*, Phys. Rev. Lett. **88**, 091802 (2002) [Erratum-ibid. **90**, 239902 (2003)] [arXiv:hep-ex/0110059].

[229] S. Davidson, S. Forte, P. Gambino, N. Rius and A. Strumia, *Old and new*

*physics interpretations of the NuTeV anomaly*, JHEP **0202** (2002) 037 [arXiv:hep-ph/0112302].

[230] W. Loinaz, N. Okamura, S. Rayyan, T. Takeuchi and L. C. R. Wijewardhana, *The NuTeV anomaly, lepton universality, and non-universal neutrino-gauge couplings*, Phys. Rev. D **70** (2004) 113004 [arXiv:hep-ph/0403306].

[231] G. A. Miller and A. W. Thomas, *Shadowing corrections and the precise determination of electroweak parameters in neutrino-nucleon scattering*, Int. J. Mod. Phys. A **20**, 95 (2005) [arXiv:hep-ex/0204007].

[232] C. Giunti and M. Laveder, *nu/e – nu/s oscillations with large neutrino mass in NuTeV?*, arXiv:hep-ph/0202152.

[233] J. F. Beacom and M. R. Vagins, *GADZOOKS! Antineutrino spectroscopy with large water Cherenkov detectors*, Phys. Rev. Lett. **93**, 171101 (2004) [arXiv:hep-ph/0309300].

[234] J. P. Deutsch, P. C. Macq and L. Van Elmbt, *Status Of Our Evidence For Conserved Vector Current: A Remark On The Measurement Of Weak Magnetism*, Phys. Rev. C **15**, 1587 (1977).

[235] Y. K. Lee Y, L. W. Mo and C. S. Wu *Experimental test of the conserved vector current theory on the beta spectra of $B^{12}$ and $N^{12}$* Phys. Rev. Lett. **10** (1963) 253.

[236] C. S. Wu *The universal Fermi interaction and the conserved vector current in beta decay* Rev. Mod. Phys. **36** (1964) 618.

[237] T. D. Lee and C. S. Wu, *Weak Interactions*, Ann. Rev. Nucl. Part. Sci. **15**, 381 (1965).

[238] Z. Daraktchieva *et al.* [MUNU Collaboration], *Final results on the neutrino magnetic moment from the MUNU experiment*, Phys. Lett. B **615** (2005) 153 [arXiv:hep-ex/0502037].

[239] H. B. Li *et al.* [TEXONO Collaboration], *New limits on neutrino magnetic moments from the Kuo-Sheng reactor neutrino experiment*, Phys. Rev. Lett. **90** (2003) 131802 [arXiv:hep-ex/0212003].

[240] F. Reines, H. S. Gurr and H. W. Sobel, *Detection Of Anti-Electron-Neutrino E Scattering*, Phys. Rev. Lett. **37** (1976) 315.

[241] H. T. Wong *et al.* [TEXONO Collaboration], *A search of neutrino magnetic moments with a high-purity germanium detector at the Kuo-Sheng nuclear power station*, Phys. Rev. D **75** (2007) 012001 [arXiv:hep-ex/0605006].

[242] A. G. Beda *et al.*, *The first result of the neutrino magnetic moment measurement in the GEMMA experiment*, arXiv:0705.4576 [hep-ex].

[243] G. G. Raffelt *Stars As Laboratories For Fundamental Physics: The Astrophysics Of Neutrinos, Axions, And Other Weakly Interacting Particles*, Chicago, USA Univ. Pr.(1996) and references therein.

# Index